Principles of Proteomics

Principles of Proteomics

Peter Wyatt

R CALLISTO REFERENCE

www.callistoreference.com

Callisto Reference,
118-35 Queens Blvd., Suite 400,
Forest Hills, NY 11375, USA

Visit us on the World Wide Web at:
www.callistoreference.com

ISBN: 978-1-64116-554-9 (Hardback)

Cataloging-in-Publication Data

Principles of proteomics / Peter Wyatt.
    p. cm.
Includes bibliographical references and index.
ISBN 978-1-64116-554-9
1. Proteomics. 2. Proteins. 3. Molecular biology. I. Wyatt, Peter.
QP551 .P75 2022
572.6--dc23

Table of Contents

Preface   **VII**

Chapter 1   **Understanding Protein and Proteomics**   **1**
- Protein   1
- Proteome   62
- Proteomics   66
- Phosphoproteomics   83

Chapter 2   **Protein Purification**   **86**
- Methods of Protein Purification   99
- Primary Structure Determination   134
- 3D Structure Determination   135

Chapter 3   **Protein-Protein Interactions: Methods for Detection and Analysis**   **157**
- Protein-Protein Interactions   157
- In Vitro Techniques   160
- In Vivo Techniques   178
- Crosslinking Protein Interaction Analysis   184
- Interactome   190

Chapter 4   **Protein Post-Translational Modifications**   **193**
- Ubiquitination   194
- Phosphorylation   195
- Glycosylation   206
- S-Nitrosylation   217
- Methylation   218
- Acetylation   220
- Lipidation   221
- Proteolysis   223

Chapter 5   **Applications of Proteomics**   **226**
- Proteomics in Biotechnology   226
- Environmental Applications of Proteomics   227
- Application of Proteomics in Food Industry   231
- Medical Application of Proteomics   232

Permissions

Index

Preface

The large-scale study of proteins is known as proteomics. A set of proteins that is produced and modified by an organism or system is named as proteome. Proteomics has led to the identification of a large number of proteins. It is involved in the discovery of proteomes from various levels of protein composition, structure and activities. Proteomics plays an important role in functional genomics as well. Some of the common methods used in this field are protein detection with antibodies and antibody-free protein detection. It is applied in numerous fields such as identifying potential drugs and for revealing complex plant-insect interactions. This textbook presents the complex subject of proteomics in the most comprehensible and easy to understand language. Some of the diverse topics covered herein address the varied branches that fall under this category. Through this book, we attempt to further enlighten the readers about the new concepts in this field.

A foreword of all Chapters of the book is provided below:

Chapter 1 - The large biomolecules that consist of one or more long chains of amino acid residues is referred to as proteins. Proteins perform various functions within organisms such as responding to stimuli, catalysing metabolic reactions, DNA replication, etc. The large scale study of proteins is known as proteomics. This is an introductory chapter which will introduce briefly all these significant aspects of proteins.; **Chapter 2** - Protein purification is the sequence of processes that isolate one or a few proteins from a complex mixture. It plays a major role in characterizing the function, structure and interactions of the protein. This chapter discusses in detail the theories related to protein purification along with the different methods used to purify proteins.; **Chapter 3** - Protein–protein interactions (PPIs) are the physical contacts of high specificity which are established between two or more protein molecules. They are caused by biochemical events caused by electrostatic forces including the hydrophobic effect. The chapter closely examines these key concepts of protein-protein interactions to provide an extensive understanding of the subject.; **Chapter 4** - Post-translational modification is the enzymatic and covalent modification of proteins that follows protein biosynthesis. Proteins are synthesized by ribosomes and then undergo post-translational modification in order to form mature protein product. This chapter has been carefully written to provide an easy understanding of the varied kinds of post-translational modifications such as acetylation, methylation and ubiquitination.; **Chapter 5** - Proteomics involves the large-scale experimental analysis of proteins and proteomes. It has helped in the identification of a large number of proteins. Proteomics is applied in a variety of fields such as biotechnology, environment, medicine, food industry etc. The diverse applications of proteomics in these different areas has been thoroughly discussed in this chapter.;

I would like to thank the entire editorial team who made sincere efforts for this book and my family who supported me in my efforts of working on this book. I take this opportunity to thank all those who have been a guiding force throughout my life.

Peter Wyatt

Chapter 1

Understanding Protein and Proteomics

The large biomolecules that consist of one or more long chains of amino acid residues is referred to as proteins. Proteins perform various functions within organisms such as responding to stimuli, catalysing metabolic reactions, DNA replication, etc. The large scale study of proteins is known as proteomics. This is an introductory chapter which will introduce briefly all these significant aspects of proteins.

Protein

Protein is a highly complex substance that is present in all living organisms. Proteins are of great nutritional value and are directly involved in the chemical processes essential for life. The importance of proteins was recognized by chemists in the early 19th century, including Swedish chemist Jöns Jacob Berzelius, who in 1838 coined the term protein, a word derived from the Greek proteios, meaning "holding first place." Proteins are species-specific; that is, the proteins of one species differ from those of another species. They are also organ-specific; for instance, within a single organism, muscle proteins differ from those of the brain and liver.

A protein molecule is very large compared with molecules of sugar or salt and consists of many amino acids joined together to form long chains, much as beads are arranged on a string. There are about 20 different amino acids that occur naturally in proteins. Proteins of similar function have similar amino acid composition and sequence. Although it is not yet possible to explain all of the functions of a protein from its amino acid sequence, established correlations between structure and function can be attributed to the properties of the amino acids that compose proteins.

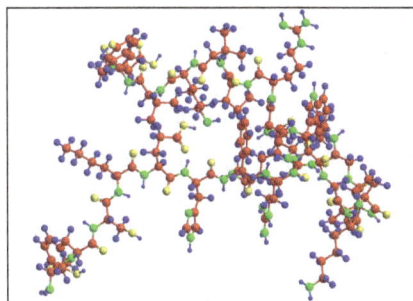

Peptide: The molecular structure of a peptide (a small protein) consists of a sequence of amino acids.

Plants can synthesize all of the amino acids; animals cannot, even though all of them are essential for life. Plants can grow in a medium containing inorganic nutrients that provide nitrogen, potassium, and other substances essential for growth. They utilize the carbon dioxide in the air during the process of photosynthesis to form organic compounds such as carbohydrates. Animals, however, must obtain organic nutrients from outside sources. Because the protein content of most plants is low, very large amounts of plant material are required by animals, such as ruminants (e.g., cows), that eat only plant material to meet their amino acid requirements. Nonruminant animals, including humans, obtain proteins principally from animals and their products—e.g., meat, milk, and eggs. The seeds of legumes are increasingly being used to prepare inexpensive protein-rich food.

The protein content of animal organs is usually much higher than that of the blood plasma. Muscles, for example, contain about 30 percent protein, the liver 20 to 30 percent, and red blood cells 30 percent. Higher percentages of protein are found in hair, bones, and other organs and tissues with a low water content. The quantity of free amino acids and peptides in animals is much smaller than the amount of protein; protein molecules are produced in cells by the stepwise alignment of amino acids and are released into the body fluids only after synthesis is complete.

The high protein content of some organs does not mean that the importance of proteins is related to their amount in an organism or tissue; on the contrary, some of the most important proteins, such as enzymes and hormones, occur in extremely small amounts. The importance of proteins is related principally to their function. All enzymes identified thus far are proteins. Enzymes, which are the catalysts of all metabolic reactions, enable an organism to build up the chemical substances necessary for life—proteins, nucleic acids, carbohydrates, and lipids—to convert them into other substances, and to degrade them. Life without enzymes is not possible. There are several protein hormones with important regulatory functions. In all vertebrates, the respiratory protein hemoglobin acts as oxygen carrier in the blood, transporting oxygen from the lung to body organs and tissues. A large group of structural proteins maintains and protects the structure of the animal body.

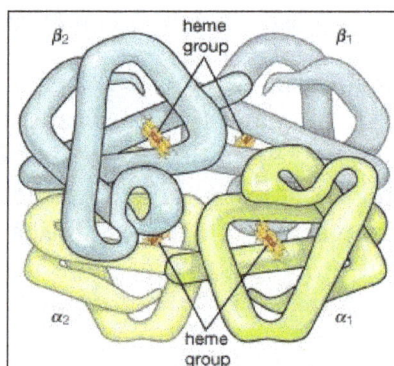

Hemoglobin is a protein made up of four polypeptide chains (α_1, α_2, β_1, and β_2). Each chain is attached to a heme group composed of porphyrin (an organic ringlike compound) attached to an iron atom. These iron-porphyrin complexes coordinate oxygen molecules reversibly, an ability directly related to the role of hemoglobin in oxygen transport in the blood.

General Structure and Properties of Proteins

The Amino Acid Composition of Proteins

The common property of all proteins is that they consist of long chains of α-amino (alpha amino) acids. The general structure of α-amino acids is shown in. The α-amino acids are so called because the α-carbon atom in the molecule carries an amino group ($-NH_2$); the α-carbon atom also carries a carboxyl group ($-COOH$).

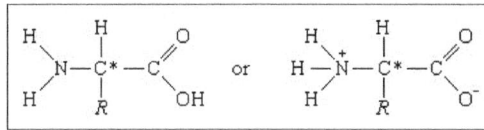

In acidic solutions, when the pH is less than 4, the $-COO$ groups combine with hydrogen ions (H^+) and are thus converted into the uncharged form ($-COOH$). In alkaline solutions, at pH above 9, the ammonium groups ($-NH^+_3$) lose a hydrogen ionand are converted into amino groups ($-NH_2$). In the pH range between 4 and 8, amino acids carry both a positive and a negative charge and therefore do not migrate in an electrical field. Such structures have been designated as dipolar ions, or zwitterions (i.e., hybrid ions).

Although more than 100 amino acids occur in nature, particularly in plants, only 20 types are commonly found in most proteins. In protein molecules the α-amino acids are linked to each other by peptide bonds between the amino group of one amino acid and the carboxyl group of its neighbour.

The condensation (joining) of three amino acids yields the tripeptide.

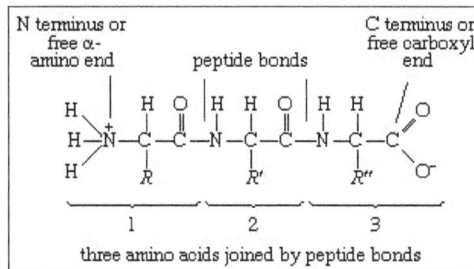

three amino acids joined by peptide bonds

It is customary to write the structure of peptides in such a way that the free α-amino group (also called the N terminus of the peptide) is at the left side and the free carboxyl group (the C terminus) at the right side. Proteins are macromolecular polypeptides—i.e., very large molecules composed of many peptide-bonded amino acids. Most of the common ones contain more than 100 amino acids linked to each other in a long peptide chain. The average molecular weight (based on the weight of a hydrogen atom as 1) of each amino acid is approximately 100 to 125; thus, the molecular weights of proteins are usually in the range of 10,000 to 100,000 daltons (one dalton is the weight of one hydrogen atom). The species-specificity and organ-specificity of proteins result from differences in the number and sequences of amino acids. Twenty different amino acids in a chain 100 amino acids long can be arranged in far more than 10^{100} ways (10^{100} is the number one followed by 100 zeroes).

Structures of Common Amino Acids

The amino acids present in proteins differ from each other in the structure of their side (R) chains. The simplest amino acid is glycine, in which R is a hydrogen atom. In a number of amino acids, R represents straight or branched carbon chains. One of these amino acids is alanine, in which R is the methyl group ($-CH_3$). Valine, leucine, and isoleucine, with longer R groups, complete the alkyl side-chain series. The alkyl side chains (R groups) of these amino acids are nonpolar; this means that they have no affinity for water but some affinity for each other. Although plants can form all of the alkyl amino acids, animals can synthesize only alanine and glycine; thus valine, leucine, and isoleucine must be supplied in the diet.

Two amino acids, each containing three carbon atoms, are derived from alanine; they are serine and cysteine. Serine contains an alcohol group ($-CH_2OH$) instead of the methyl group of alanine, and cysteine contains a mercapto group ($-CH_2SH$). Animals can synthesize serine but not cysteine or cystine. Cysteine occurs in proteins predominantly in its oxidized form (oxidation in this sense meaning the removal of hydrogen atoms), called cystine. Cystine consists of two cysteine molecules linked by the disulfide bond ($-S-S-$) that results when a hydrogen atom is removed from the mercapto group of each of the cysteines. Disulfide bonds are important in protein structure because they allow the linkage of two different parts of a protein molecule to—and thus the formation of loops in—the otherwise straight chains. Some proteins contain small amounts of cysteine with free sulfhydryl ($-SH$) groups.

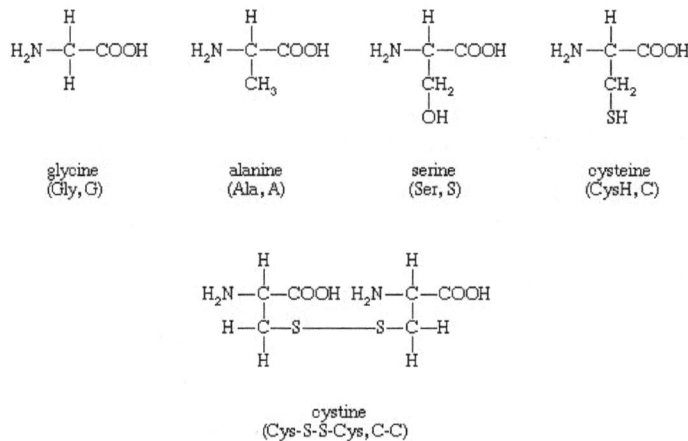

glycine
(Gly, G)

alanine
(Ala, A)

serine
(Ser, S)

cysteine
(CysH, C)

cystine
(Cys-S-S-Cys, C-C)

Four amino acids, each consisting of four carbon atoms, occur in proteins; they are aspartic acid, asparagine, threonine, and methionine. Aspartic acid and asparagine, which occur in large amounts, can be synthesized by animals. Threonine and methionine cannot be synthesized and thus are essential amino acids; i.e., they must be supplied in the diet. Most proteins contain only small amounts of methionine.

Proteins also contain an amino acid with five carbon atoms (glutamic acid) and a secondary amine (in proline), which is a structure with the amino group ($-NH_2$) bonded to the alkyl side chain, forming a ring. Glutamic acid and aspartic acid are dicarboxylic acids; that is, they have two carboxyl groups ($-COOH$).

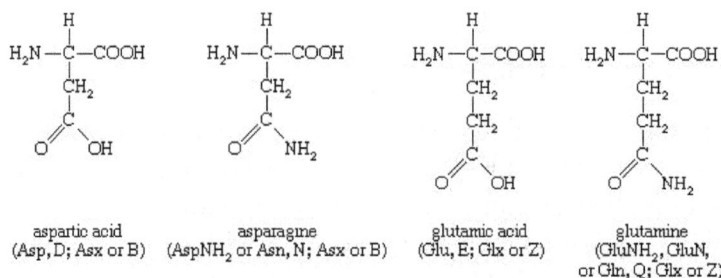

aspartic acid
(Asp, D; Asx or B)

asparagine
(AspNH$_2$ or Asn, N; Asx or B)

glutamic acid
(Glu, E; Glx or Z)

glutamine
(GluNH$_2$, GluN,
or Gln, Q; Glx or Z)

Glutamine is similar to asparagine in that both are the amides of their corresponding dicarboxylic acid forms; i.e., they have an amide group ($-CONH_2$) in place of the carboxyl ($-COOH$) of the side chain. Glutamic acid and glutamine are abundant in most proteins; e.g., in plant proteins they sometimes comprise more than one-third of the amino acids present. Both glutamic acid and glutamine can be synthesized by animals.

Amino acid	Amino acid content of some proteins*					
	protein					
	Alpha-casein	Gliadin	Edestin	Collagen (ox hide)	Keratin (wool)	Myosin
Lysine	60.9	4.45	19.9	27.4	6.2	85
Histidine	18.7	11.7	18.6	4.5	19.7	15
Arginine	24.7	15.7	99.2	47.1	56.9	41
Aspartic acid**	63.1	10.1	99.4	51.9	51.5	85
Threonine	41.2	17.6	31.2	19.3	55.9	41
Serine	63.1	46.7	55.7	41.0	79.5	41
Glutamic acid**	153.1	311.0	144.9	76.2	99.0	155
Proline	71.3	117.8	32.9	125.2	58.3	22
Glycine	37.3	-	68.0	354.6	78.0	39
Alanine	41.5	23.9	57.7	115.7	43.8	78
Half-cystine	3.6	21.3	10.9	0.0	105.0	86
Valine	53.8	22.7	54.6	21.4	46.6	42
Methionine	16.8	11.3	16.4	6.5	4.0	22
Isoleucine	48.8	90.8**	41.9	14.5	29.0	42
Leucine	60.3		60.0	28.2	59.9	79
Tyrosine	44.7	17.7	26.9	5.5	28.7	18
Phenylalanine	27.9	39.0	38.1	13.9	22.4	27
Tryptophan	7.8	3.2	6.6	0.0	9.6	-
Hydroxyproline	0.0	0.0	0.0	97.5	12.2	-
Hydroxylysine	-	-	-	8.0	1.2	-
Total	839	765	883	1,058	863	832
Average residual weight	119	131	113	95	117	120

*Number of gram molecules of amino acid per 100,000 grams of protein.

**The values for aspartic acid and glutamic acid include asparagine and glutamine, respectively.

***Isoleucine plus leucine.

The amino acids proline and hydroxyproline occur in large amounts in collagen, the protein of the connective tissue of animals. Proline and hydroxyproline lack free amino ($-NH_2$) groups because the amino group is enclosed in a ring structure with the side chain; they thus cannot exist in a zwitterion form. Although the nitrogen-containing group ($>NH$) of these amino acids can form a peptide bond with the carboxyl group of another amino acid, the bond so formed gives rise to a kink in the peptide chain; i.e., the ring structure alters the regular bond angle of normal peptide bonds.

Proteins usually are almost neutral molecules; that is, they have neither acidic nor basic properties. This means that the acidic carboxyl ($-COO^-$) groups of aspartic and glutamic acid are about equal in number to the amino acids with basic side chains. Three such basic amino acids, each containing six carbon atoms, occur in proteins. The one with the simplest structure, lysine, is synthesized by plants but not by animals. Even some plants have a low lysine content. Arginine is found in all proteins; it occurs in particularly high amounts in the strongly basic protamines (simple proteins composed of relatively few amino acids) of fish sperm. The third basic amino acid is histidine. Both arginine and histidine can be synthesized by animals. Histidine is a weaker base than either lysine or arginine. The imidazole ring, a five-membered ring structure containing two nitrogen atoms in the side chain of histidine, acts as a buffer (i.e., a stabilizer of hydrogen ion concentration) by binding hydrogen ions (H^+) to the nitrogen atoms of the imidazole ring.

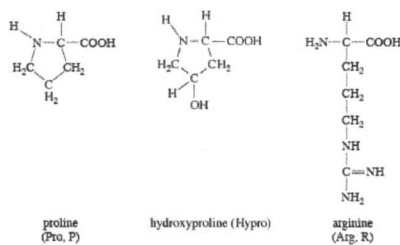

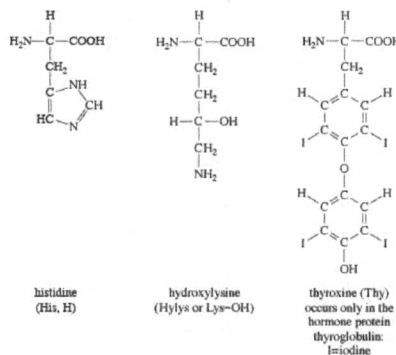

proline (Pro, P) hydroxyproline (Hypro) arginine (Arg, R) histidine (His, H) hydroxylysine (Hylys or Lys–OH) thyroxine (Thy) occurs only in the hormone protein thyroglobulin; I=iodine

The remaining amino acids—phenylalanine, tyrosine, and tryptophan—have in common an aromatic structure; i.e., a benzene ring is present. These three amino acids are essential, and, while animals cannot synthesize the benzene ring itself, they can convert phenylalanine to tyrosine.

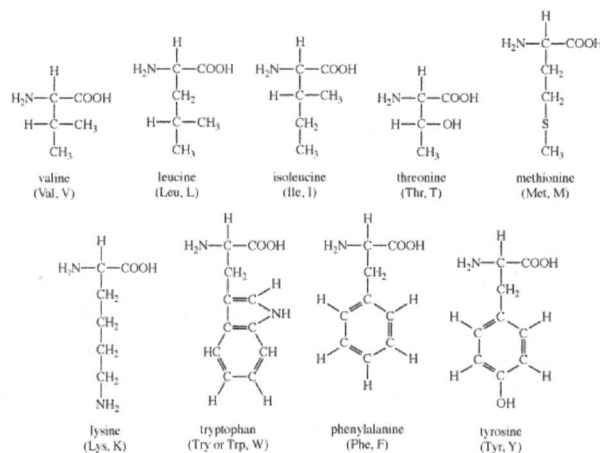

valine (Val, V) leucine (Leu, L) isoleucine (Ile, I) threonine (Thr, T) methionine (Met, M) lysine (Lys, K) tryptophan (Try or Trp, W) phenylalanine (Phe, F) tyrosine (Tyr, Y)

Because these amino acids contain benzene rings, they can absorb ultraviolet light at wavelengths between 270 and 290 nanometres (nm; 1 nanometre = 10^{-9} metre = 10 angstrom units). Phenylalanine absorbs very little ultraviolet light; tyrosine and tryptophan, however, absorb it strongly and are responsible for the absorption band most proteins exhibit at 280–290 nanometres. This absorption is often used to determine the quantity of protein present in protein samples.

Most proteins contain only the amino acids described above; however, other amino acids occur in proteins in small amounts. For example, the collagen found in connective tissue contains, in addition to hydroxyproline, small amounts of hydroxylysine. Other proteins contain some monomethyl-, dimethyl-, or trimethyllysine—i.e., lysine derivatives containing one, two, or three methyl groups ($-CH_3$). The amount of these unusual amino acids in proteins, however, rarely exceeds 1 or 2 percent of the total amino acids.

Physicochemical Properties of the Amino Acids

The physicochemical properties of a protein are determined by the analogous properties of the amino acids in it.

The α-carbon atom of all amino acids, with the exception of glycine, is asymmetric; this means that four different chemical entities (atoms or groups of atoms) are attached to it. As a result, each of the amino acids, except glycine, can exist in two different spatial, or geometric, arrangements (i.e., isomers), which are mirror images akin to right and left hands.

These isomers exhibit the property of optical rotation. Optical rotation is the rotation of the plane of polarized light, which is composed of light waves that vibrate in one plane, or direction, only. Solutions of substances that rotate the plane of polarization are said to be optically active, and the degree of rotation is called the optical rotation of the solution. The direction in which the light is rotated is generally designed as plus, or d, for dextrorotatory (to the right), or as minus, or l, for levorotatory (to the left). Some amino acids are dextrorotatory, others are levorotatory. With the exception of a few small proteins (peptides) that occur in bacteria, the amino acids that occur in proteins are L-amino acids.

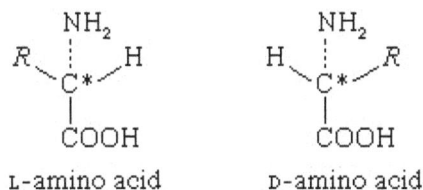

L-amino acid   D-amino acid

In bacteria, D-alanine and some other D-amino acids have been found as components of gramicidin and bacitracin. These peptides are toxic to other bacteria and are used in medicine as antibiotics. The D-alanine has also been found in some peptides of bacterial membranes.

In contrast to most organic acids and amines, the amino acids are insoluble in organic solvents. In aqueous solutions they are dipolar ions (zwitterions, or hybrid ions) that react with strong acids or bases in a way that leads to the neutralization of the negatively or positively charged ends, respectively. Because of their reactions with strong acids and strong bases, the amino acids act as buffers—stabilizers of hydrogen ion (H^+) or hydroxide ion (OH^-) concentrations. In fact, glycine is frequently used as a buffer in the pH range from 1 to 3 (acid solutions) and from 9 to 12 (basic solutions). In acid solutions, glycine has a positive charge and therefore migrates to the cathode(negative electrode of a direct-current electrical circuit with terminals in the solution). Its charge, however, is negative in alkaline solutions, in which it migrates to the anode(positive electrode). At pH 6.1 glycine does not migrate, because each molecule has one positive and one negative charge. The pH at which an amino acid does not migrate in an electrical field is called the isoelectric point. Most of the monoamino acids (i.e., those with only one amino group) have isoelectric points similar to that of glycine. The isoelectric points of aspartic and glutamic acids, however, are close to pH 3, and those of histidine, lysine, and arginine are at pH 7.6, 9.7, and 10.8, respectively.

Amino Acid Sequence in Protein Molecules

Since each protein molecule consists of a long chain of amino acid residues, linked to each other by peptide bonds, the hydrolytic cleavage of all peptide bonds is a prerequisite for the quantitative determination of the amino acid residues. Hydrolysisis most frequently accomplished by boiling the protein with concentrated hydrochloric acid. The quantitative determination of the amino acids is based on the discovery that amino acids can be separated from each other by chromatography on filter paper and made visible by spraying the paper with ninhydrin. The amino acids of the protein hydrolysate are separated from each other by passing the hydrolysate through a column of adsorbents, which adsorb the amino acids with different affinities and, on washing the column with buffer solutions, release them in a definite order. The amount of each of the amino acids can be determined by the intensity of the colour reaction with ninhydrin.

To obtain information about the sequence of the amino acid residues in the protein, the protein is degraded stepwise, one amino acid being split off in each step. This is accomplished by coupling the free α-amino group ($-NH_2$) of the N-terminal amino acid with phenyl isothiocyanate; subsequent mild hydrolysis does not affect the peptide bonds. The procedure, called the Edman degradation, can be applied repeatedly; it thus reveals the sequence of the amino acids in the peptide chain.

Unavoidable small losses that occur during each step make it impossible to determine the sequence of more than about 30 to 50 amino acids by this procedure. For this reason the protein is usually first hydrolyzed by exposure to the enzyme trypsin, which cleaves only peptide bonds formed by the carboxyl groups of lysine and arginine. The Edman degradation is then applied to each of the few resulting peptides produced by the action of trypsin. Further information can be gained by hydrolyzing another portion of the protein with another enzyme, for instance with chymotrypsin, which splits predominantly peptide bonds formed by the amino acids tyrosine, phenylalanine, and tryptophan. The combination of results obtained with two or more different proteolytic (protein degrading) enzymes was first applied by English biochemist Frederick Sanger, and it enabled him to elucidate the amino acid sequence of insulin. The amino acid sequences of many other proteins subsequently were determined in the same manner.

Levels of Structural Organization in Proteins

Primary Structure

Analytical and synthetic procedures reveal only the primary structure of the proteins—that is, the amino acid sequence of the peptide chains. They do not reveal information about the conformation (arrangement in space) of the peptide chain—that is, whether the peptide chain is present as a long straight thread or is irregularly coiled and folded into a globule. The configuration, or conformation, of a protein is determined by mutual attraction or repulsion of polar or nonpolar groups in the side chains (R groups) of the amino acids. The former have positive or negative charges in their side chains; the latter repel water but attract each other. Some parts of a peptide chain containing 100 to 200 amino acids may form a loop, or helix; others may be straight or form irregular coils.

The terms secondary, tertiary, and quaternary structure are frequently applied to the configuration of the peptide chain of a protein. A nomenclature committee of the International Union of Biochemistry (IUB) has defined these terms as follows: The primary structure of a protein is determined by its amino acid sequence without any regard for the arrangement of the peptide chain in space. The secondary structure is determined by the spatial arrangement of the main peptide chain without any regard for the conformation of side chains or other segments of the main chain. The tertiary structure is determined by both the side chains and other adjacent segments of the main chain, without regard for neighbouring peptide chains. Finally, the term quaternary structure is used for the arrangement of identical or different subunits of a large protein in which each subunit is a separate peptide chain.

Secondary Structure

The nitrogen and carbon atoms of a peptide chain cannot lie on a straight line, because of the magnitude of the bond angles between adjacent atoms of the chain; the bond angle is about 110°. Each of the nitrogen and carbon atoms can rotate to a certain extent, however, so that the chain has a limited flexibility. Because all of the amino acids, except glycine, are asymmetric L-amino acids, the peptide chain tends to assume an asymmetric helical shape; some of the fibrous proteins consist of elongated helices around a straight screw axis. Such structural features result from properties common to all peptide chains. The product of their effects is the secondary structure of the protein.

Tertiary Structure

The tertiary structure is the product of the interaction between the side chains (R) of the amino acids composing the protein. Some of them contain positively or negatively charged groups, others are polar, and still others are nonpolar. The number of carbon atoms in the side chain varies from zero in glycine to nine in tryptophan. Positively and negatively charged side chains have the tendency to attract each other; side chains with identical charges repel each other. The bonds formed by the forces between the negatively charged side chains of aspartic or glutamic acid on the one hand, and the positively charged side chains of lysine or arginine on the other hand, are called salt bridges. Mutual attraction of adjacent peptide chains also results from the formation of numerous hydrogen bonds.

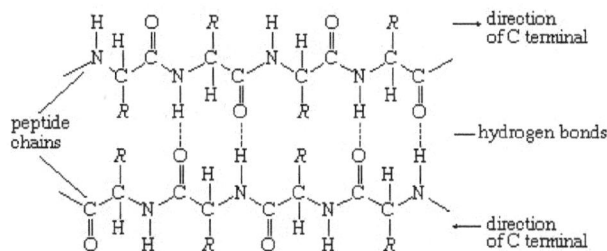

Hydrogen bonds form as a result of the attraction between the nitrogen-bound hydrogen atom (the imide hydrogen) and the unshared pair of electrons of the oxygen atom in the double bonded carbon–oxygen group (the carbonyl group). The result is a slight displacement of the imide hydrogen toward the oxygen atom of the carbonyl group. Although the hydrogen bond is much weaker than a covalent bond (i.e., the type of bond between two carbon atoms, which equally share the pair of bonding electrons between them), the large number of imide and carbonyl groups in peptide chains results in the formation of numerous hydrogen bonds. Another type of attraction is that between nonpolar side chains of valine, leucine, isoleucine, and phenylalanine; the attraction results in the displacement of water molecules and is called hydrophobic interaction.

In proteins rich in cystine, the conformation of the peptide chain is determined to a considerable extent by the disulfide bonds ($-S-S-$) of cystine. The halves of cystine may be located in different parts of the peptide chain and thus may form a loop closed by the disulfide bond.

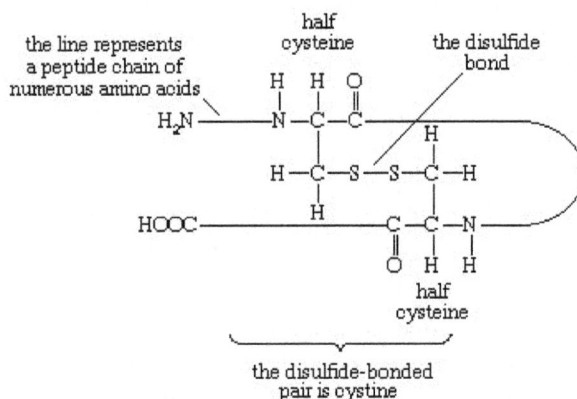

If the disulfide bond is reduced (i.e., hydrogen is added) to two sulfhydryl ($-SH$) groups, the tertiary structure of the protein undergoes a drastic change—closed loops are broken and adjacent disulfide-bonded peptide chains separate.

Quaternary Structure

The nature of the quaternary structure is demonstrated by the structure of hemoglobin. Each molecule of human hemoglobin consists of four peptide chains, two α-chains and two β-chains; i.e., it is a tetramer. The four subunits are linked to each other by hydrogen bonds and hydrophobic interaction. Because the four subunits are so closely linked, the hemoglobin tetramer is called a molecule, even though no covalent bonds occur between the peptide chains of the four subunits. In other proteins, the subunits are bound to each other by covalent bonds (disulfide bridges).

The amino acid sequence of porcine proinsulin is shown below. The arrows indicate the direction from the N terminus of the β-chain (B) to the C terminus of the α-chain (A).

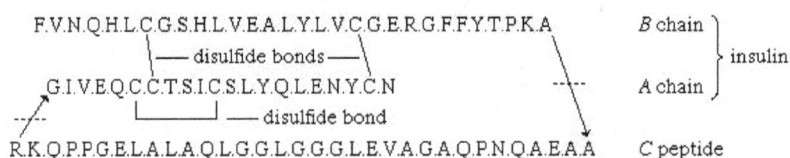

The Isolation and Determination of Proteins

Animal material usually contains large amounts of protein and lipids and small amounts of carbohydrate; in plants, the bulk of the dry matter is usually carbohydrate. If it is necessary to determine the amount of protein in a mixture of animal foodstuffs, a sample is converted to ammonium salts by boiling with sulfuric acid and a suitable inorganic catalyst, such as copper sulfate (Kjeldahl method). The method is based on the assumption that proteins contain 16 percent nitrogen, and that nonprotein nitrogen is present in very small amounts. The assumption is justified for most tissues from higher animals but not for insects and crustaceans, in which a considerable portion of the body nitrogen is present in the form of chitin, a carbohydrate. Large amounts of nonprotein nitrogen are also found in the sap of many plants. In such cases, the precise quantitative analyses are made after the proteins have been separated from other biological compounds.

Proteins are sensitive to heat, acids, bases, organic solvents, and radiation exposure; for this reason, the chemical methods employed to purify organic compounds cannot be applied to proteins. Salts and molecules of small size are removed from protein solutions by dialysis—i.e., by placing the solution into a sac of semipermeable material, such as cellulose or acetylcellulose, which will allow small molecules to pass through but not large protein molecules, and immersing the sac in water or a salt solution. Small molecules can also be removed either by passing the protein solution through a column of resin that adsorbs only the protein or by gel filtration. In gel filtration, the large protein molecules pass through the column, and the small molecules are adsorbed to the gel.

Groups of proteins are separated from each other by salting out—i.e., the stepwise addition of sodium sulfate or ammonium sulfate to a protein solution. Some proteins, called globulins, become insoluble and precipitate when the solution is half-saturated with ammonium sulfate or when its sodium sulfate content exceeds about 12 percent. Other proteins, the albumins, can be precipitated from the supernatant solution (i.e., the solution remaining after a precipitation has taken place) by saturation with ammonium sulfate. Water-soluble proteins can be obtained in a dry state by freeze-drying (lyophilization), in which the protein solution is deep-frozen by lowering the temperature below −15 °C (5 °F) and removing the water; the protein is obtained as a dry powder.

Most proteins are insoluble in boiling water and are denatured by it—i.e., irreversibly converted into an insoluble material. Heat denaturation cannot be used with connective tissue because the principal structural protein, collagen, is converted by boiling water into water-soluble gelatin.

Fractionation (separation into components) of a mixture of proteins of different molecular weight can be accomplished by gel filtration. The size of the proteins retained by the gel depends upon the properties of the gel. The proteins retained in the gel are removed from the column by solutions of a suitable concentration of salts and hydrogen ions.

Many proteins were originally obtained in crystalline form, but crystallinity is not proof of purity; many crystalline protein preparations contain other substances. Various tests are used to determine whether a protein preparation contains only one protein. The purity of a protein solution can be determined by such techniques as chromatography and gel filtration. In addition, a solution of pure protein will yield one peak when spun in a centrifuge at very high speeds (ultracentrifugation) and will migrate as a single band in electrophoresis (migration of the protein in an electrical field). After these methods and others (such as amino acid analysis) indicate that the protein solution is pure, it can be considered so. Because chromatography, ultracentrifugation, and electrophoresis cannot be applied to insoluble proteins, little is known about them; they may be mixtures of many similar proteins.

Very small (microheterogeneous) differences in some of the apparently pure proteins are known to occur. They are differences in the amino acid composition of otherwise identical proteins and are transmitted from generation to generation; i.e., they are genetically determined. For example, some humans have two hemoglobins, hemoglobin A and hemoglobin S, which differ in one amino acid at a specific site in the molecule. In hemoglobin A the site is occupied by glutamic acid and in hemoglobin S by valine. Refinement of the techniques of protein analysis has resulted in the discovery of other instances of microheterogeneity.

The quantity of a pure protein can be determined by weighing or by measuring the ultraviolet absorbancy at 280 nanometres. The absorbency at 280 nanometres depends on the content of tyrosine and tryptophan in the protein. Sometimes the slightly less sensitive biuret reaction, a purple colour given by alkaline protein solutions upon the addition of copper sulfate, is used; its intensity depends only on the number of peptide bonds per gram, which is similar in all proteins.

Physicochemical Properties of Proteins

The Molecular Weight of Proteins

The molecular weight of proteins cannot be determined by the methods of classical chemistry (e.g., freezing-point depression), because they require solutions of a higher concentration of protein than can be prepared.

If a protein contains only one molecule of one of the amino acids or one atom of iron, copper, or another element, the minimum molecular weight of the protein or a subunit can be calculated; for example, the protein myoglobin contains 0.34 gram of iron in 100 grams of protein. The atomic weight of iron is 56; thus the minimum molecular weight of myoglobin is $(56 \times 100)/0.34 =$ about 16,500. Direct measurements of the molecular weight of myoglobin yield the same value. The molecular weight of hemoglobin, however, which also contains 0.34 percent iron, has been found to be 66,000 or $4 \times 16,500$; thus hemoglobin contains four atoms of iron.

The method most frequently used to determine the molecular weight of proteins is ultracentrifugation—i.e., spinning in a centrifuge at velocities up to about 60,000 revolutions per minute. Centrifugal forces of more than 200,000 times the gravitational force on the surface of Earth are achieved at such velocities. The first ultracentrifuges, built in 1920, were used to determine the molecular weight of proteins. The molecular weights of a large number of proteins have been determined. Most consist of several subunits, the molecular weight of which is usually less than 100,000 and frequently ranges from 20,000 to 30,000. Proteins of very high molecular weights are found among hemocyanins, the copper-containing respiratory proteins of invertebrates; some range as high as several million. Although there is no definite lower limit for the molecular weight of proteins, short amino acid sequences are usually called peptides.

The Shape of Protein Molecules

In the technique of X-ray diffraction, the X-rays are allowed to strike a protein crystal. The X-rays, diffracted (bent) by the crystal, impinge on a photographic plate, forming a pattern of spots. This method reveals that peptide chains can assume very complicated, apparently irregular shapes. Two extremes in shape include the closely folded structure of the globular proteins and the elongated, unidimensional structure of the threadlike fibrous proteins; both were recognized many years before the technique of X-ray diffraction was developed. Solutions of fibrous proteins are extremely viscous (i.e., sticky); those of the globular proteins have low viscosity (i.e., they flow easily). A 5 percent solution of a globular protein—ovalbumin, for example—easily flows through a narrow glass tube; a 5 percent solution of gelatin, a fibrous protein, however, does not flow through the tube, because it is liquid only at high temperatures and solidifies at room temperature. Even solutions containing only 1 or 2 percent of gelatin are highly viscous and flow through a narrow tube either very slowly or only under pressure.

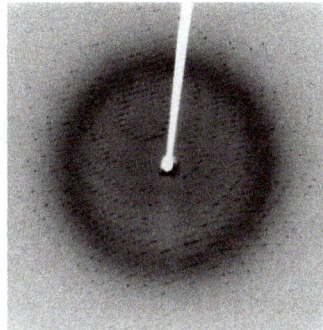

X-ray diffraction pattern of a crystallized enzyme.

The elongated peptide chains of the fibrous proteins can be imagined to become entangled not only mechanically but also by mutual attraction of their side chains, and in this way they incorporate large amounts of water. Most of the hydrophilic (water-attracting) groups of the globular proteins, however, lie on the surface of the molecules, and, as a result, globular proteins incorporate only a few water molecules. If a solution of a fibrous protein flows through a narrow tube, the elongated molecules become oriented parallel to the direction of the flow, and the solution thus becomes birefringent like a crystal; i.e., it splits a light ray into two components that travel at different velocities and are polarized at right angles to each other. Globular proteins do not show this phenomenon, which is called flow birefringence. Solutions of myosin, the contractile protein of muscles, show very high flow birefringence; other proteins with very high flow birefringence include solutions of fibrinogen, the clotting material of blood plasma, and solutions of tobacco mosaic virus. The gamma-globulins of the blood plasma show low flow birefringence, and none can be observed in solutions of serum albumin and ovalbumin.

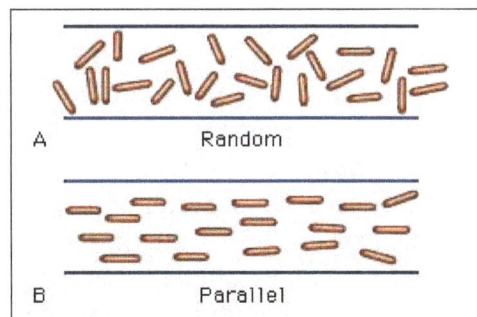

Flow birefringence. Orientation of elongated, rodlike macromolecules
(A) in resting solution, or (B) during flow through a horizontal tube.

Hydration of Proteins

When dry proteins are exposed to air of high water content, they rapidly bind water up to a maximum quantity, which differs for different proteins; usually it is 10 to 20 percent of the weight of the protein. The hydrophilic groups of a protein are chiefly the positively charged groups in the side chains of lysine and arginine and the negatively charged groups of aspartic and glutamic acid. Hydration (i.e., the binding of water) may also occur at the hydroxyl ($-OH$) groups of serine and threonine or at the amide ($-CONH_2$) groups of asparagine and glutamine.

The binding of water molecules to either charged or polar (partly charged) groups is explained by the dipolar structure of the water molecule; that is, the two positively charged hydrogen atoms form an angle of about 105°, with the negatively charged oxygen atom at the apex. The centre of the positive charges is located between the two hydrogen atoms; the centre of the negative charge of the oxygen atom is at the apex of the angle. The negative pole of the dipolar water molecule binds to positively charged groups; the positive pole binds negatively charged ones. The negative pole of the water molecule also binds to the hydroxyl and amino groups of the protein.

The water of hydration is essential to the structure of protein crystals; when they are completely dehydrated, the crystalline structure disintegrates. In some proteins this process is accompanied by denaturation and loss of the biological function.

In aqueous solutions, proteins bind some of the water molecules very firmly; others are either very loosely bound or form islands of water molecules between loops of folded peptide chains. Because the water molecules in such an island are thought to be oriented as in ice, which is crystalline water, the islands of water in proteins are called icebergs. Water molecules may also form bridges between the carbonyl and imino groups of adjacent peptide chains, resulting in structures similar to those of the pleated sheet but with a water molecule in the position of the hydrogen bonds of that configuration. The extent of hydration of protein molecules in aqueous solutions is important, because some of the methods used to determine the molecular weight of proteins yield the molecular weight of the hydrated protein. The amount of water bound to one gram of a globular protein in solution varies from 0.2 to 0.5 gram. Much larger amounts of water are mechanically immobilized between the elongated peptide chains of fibrous proteins; for example, one gram of gelatin can immobilize at room temperature 25 to 30 grams of water.

Hydration of proteins is necessary for their solubility in water. If the water of hydration of a protein dissolved in water is reduced by the addition of a salt such as ammonium sulfate, the protein is no longer soluble and is salted out, or precipitated. The salting-out process is reversible because the protein is not denatured (i.e., irreversibly converted to an insoluble material) by the addition of such salts as sodium chloride, sodium sulfate, or ammonium sulfate. Some globulins, called euglobulins, are insoluble in water in the absence of salts; their insolubility is attributed to the mutual interaction of polar groups on the surface of adjacent molecules, a process that results in the formation of large aggregates of molecules. Addition of small amounts of salt causes the euglobulins to become soluble. This process, called salting in, results from a combination between anions (negatively charged ions) and cations (positively charged ions) of the salt and positively and negatively charged side chains of the euglobulins. The combination prevents the aggregation of euglobulin molecules by preventing the formation of salt bridges between them. The addition of more sodium or ammonium sulfate causes the euglobulins to salt out again and to precipitate.

Electrochemistry of Proteins

Because the α-amino group and α-carboxyl group of amino acids are converted into peptide bonds in the protein molecule, there is only one α-amino group (at the Nterminus) and one α-carboxyl group (at the C terminus) in a given protein molecule. The electrochemical character of a protein is affected very little by these two groups. Of importance, however, are the numerous positively charged ammonium groups ($-NH_3^+$) of lysine and arginine and the negatively charged carboxyl groups ($-COO^-$) of aspartic acid and glutamic acid. In most proteins, the number of positively and negatively charged groups varies from 10 to 20 per 100 amino acids.

Electrometric Titration

When measured volumes of hydrochloric acid are added to a solution of protein in salt-free water, the pH decreases in proportion to the amount of hydrogen ions added until it is about 4. Further addition of acid causes much less decrease in pH because the protein acts as a buffer at pH values of 3 to 4. The reaction that takes place in this pH range is the protonation of the carboxyl group—i.e., the conversion of $-COO-$ into $-COOH$. Electrometric titration of an isoelectric protein with potassium hydroxide causes a very slow increase in pH and a weak buffering action of the protein at pH 7; a very strong buffering action occurs in the pH range from 9 to 10. The buffering action at pH 7, which is caused by loss of protons (positively charged hydrogen) from the imidazolium groups (i.e., the five-member ring structure in the side chain) of histidine, is weak because the histidine content of proteins is usually low. The much stronger buffering action at pH values from 9 to 10 is caused by the loss of protons from the hydroxyl group of tyrosine and from the ammonium groups of lysine. Finally, protons are lost from the guanidinium groups (i.e., the nitrogen-containing terminal portion of the arginine side chains) of arginine at pH 12. Electrometric titrations of proteins yield similar curves. Electrometric titration makes possible the determination of the approximate number of carboxyl groups, ammonium groups, histidines, and tyrosines per molecule of protein.

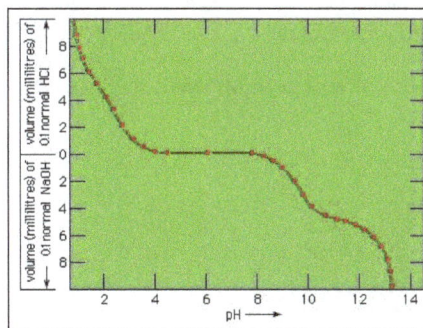

Electrometric titration of glycine.

Electrophoresis

The positively and negatively charged side chains of proteins cause them to behave like amino acids in an electrical field; that is, they migrate during electrophoresis at low pH values to the cathode (negative terminal) and at high pH values to the anode (positive terminal). The isoelectric point, the pH value at which the protein moleculedoes not migrate, is in the range of pH 5 to 7 for many proteins. Proteins such as lysozyme, cytochrome *c*, histone, and others rich in lysine and arginine, however, have isoelectric points in the pH range between 8 and 10. The isoelectric point of pepsin, which contains very few basic amino acids, is close to 1.

Number of amino acids per protein molecule							
Amino acid	Protein*						
	Cyto	Hb alpha	Hb beta	Rnase	Lys	Chgen	Fdox
Lysine	18	11	11	10	6	14	4
Histidine	3	10	9	4	1	2	1
Arginine	2	3	3	4	11	4	1
Aspartic acid**	8	12	13	15	21	23	13
Threonine	7	9	7	10	7	23	8
Serine	2	11	5	15	10	28	7
Glutamic acid**	10	5	11	12	5	15	13
Proline	4	7	7	4	2	9	4
Glycine	13	7	13	3	12	23	6
Alanine	6	21	15	12	12	22	9
Half-cystine	2	1	2	8	8	10	5
Valine	3	13	18	9	6	23	7
Methionine	3	2	1	4	2	2	0
Isoleucine	8	0	0	3	6	10	4
Leucine	6	18	18	2	8	19	8
Tyrosine	5	3	3	6	3	4	4
Phenylalanine	3	7	8	3	3	6	2
Tryptophan	1	1	2	0	6	8	1
Total	104	141	146	124	129	245	97

*Cyto = human cytochrome c; Hb alpha = human hemoglobin A, alpha-chain; Hb beta = human hemoglobin A, beta-chain; RNase = bovine ribonuclease; Lys = chicken lysozyme; Chgen = bovine chymotrypsinogen; Fdox = spinach ferredoxin.

**The values recorded for aspartic acid and glutamic acid include asparagine and glutamine, respectively.

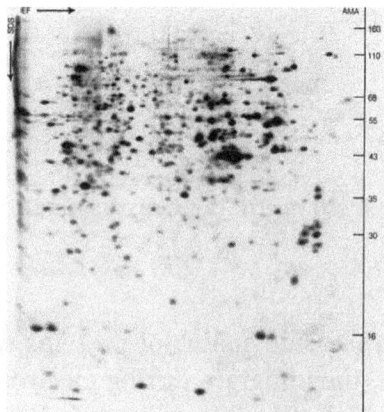

Two-dimensional gel electrophoresis. In two-dimensional gel electrophoresis, proteins are separated based on charge and size. Approaches commonly employed include isoelectric focusing (IEF) sodium dodecyl sulfate (SDS) polyacrylamide gel electrophoresis (PAGE) and immobilized pH gradient (IPG-Dalt) SDS-PAGE.

Free-boundary electrophoresis, the original method of determining electrophoretic migration, has been replaced in many instances by zone electrophoresis, in which the protein is placed in either a gel of starch, agar, or polyacrylamide or in a porous medium such as paper or cellulose acetate. The migration of hemoglobin and other coloured proteins can be followed visually. Colourless proteins are made visible after the completion of electrophoresis by staining them with a suitable dye.

Conformation of Globular Proteins

Results of X-ray Diffraction Studies

Most knowledge concerning secondary and tertiary structure of globular proteins has been obtained by the examination of their crystals using X-ray diffraction. In this technique, X-rays are allowed to strike the crystal; the X-rays are diffracted by the crystal and impinge on a photographic plate, forming a pattern of spots. The measured intensity of the diffraction pattern, as recorded on a photographic film, depends particularly on the electron density of the atoms in the protein crystal. This density is lowest in hydrogen atoms, and they do not give a visible diffraction pattern. Although carbon, oxygen, and nitrogen atoms yield visible diffraction patterns, they are present in such great number—about 700 or 800 per 100 amino acids—that the resolution of the structure of a protein containing more than 100 amino acids is almost impossible. Resolution is considerably improved by substituting into the side chains of certain amino acids very heavy atoms, particularly those of heavy metals. Mercury ions, for example, bind to the sulfhydryl ($-SH$) groups of cysteine. Platinum chloride has been used in other proteins. In the iron-containing proteins, the iron atom already in the molecule is adequate.

Although the X-ray diffraction technique cannot resolve the complete three-dimensional conformation (that is, the secondary and tertiary structure of the peptide chain), complete resolution has been obtained by combination of the results of X-ray diffraction with those of amino acid sequence analysis. In this way the complete conformation of such proteins as myoglobin, chymotrypsinogen, lysozyme, and ribonuclease has been resolved.

The X-ray diffraction method has revealed regular structural arrangements in proteins; one is an extended form of antiparallel peptide chains that are linked to each other by hydrogen bonds between the carbonyl and imino groups. This conformation, called the pleated sheet, or β-structure, is found in some fibrous proteins. Short strands of the β-structure have also been detected in some globular proteins.

A second important structural arrangement is the α-helix; it is formed by a sequence of amino acids wound around a straight axis in either a right-handed or a left-handed spiral. Each turn of the helix corresponds to a distance of 5.4 angstroms (= 0.54 nanometre) in the direction of the screw axis and contains 3.7 amino acids. Hence, the length of the α-helix per amino acid residue is 5.4 divided by 3.7, or 1.5 angstroms (1 angstrom = 0.1 nanometre). The stability of the α-helix is maintained by hydrogen bonds between the carbonyl and imino groups of neighbouring turns of the helix. It was once thought, based on data from analyses of the myoglobin molecule, more than half of which consists of α-helices, that the α-helix is the predominant structural element of the globular proteins; it is now known that myoglobin is exceptional in this respect. The other globular proteins for which the structures have been resolved by X-ray diffraction contain only small regions of α-helix. In most of them the peptide chains are folded in an apparently random fashion, for which the term random coil has been used. The term is misleading, however, because the folding is not random; rather, it is dictated by the primary structure and modified by the secondary and tertiary structures.

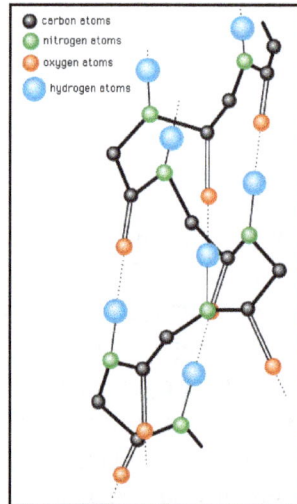

Protein structure; α-helix
The α-helix in the structural arrangement of a protein.

The first proteins for which the internal structures were completely resolved are the iron-containing proteins myoglobin and hemoglobin. The investigation of the hydrated crystals of these proteins by Austrian-born British biochemist Max Perutz and British biochemist John C. Kendrew, who won the 1962 Nobel Prize for Chemistry for their work, revealed that the folding of the peptide chains is so tight that most of the water is displaced from the centre of the globular molecules. The amino acids that carry the ammonium ($-NH_3^+$) and carboxyl ($-COO^-$) groups were found to be shifted to the surface of the globular molecules, and the nonpolar amino acids were found to be concentrated in the interior.

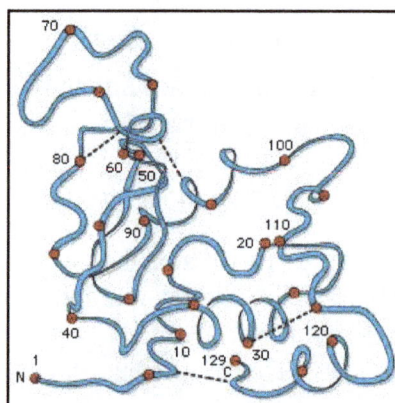

Lysozyme; protein conformation.

The simplified structure of lysozyme from hen's egg white has a single peptide chain of 129 amino acids. The amino acid residues are numbered from the terminal α group (N) to the terminal carboxyl group (C). Circles indicate every fifth residue, and every tenth residue is numbered. Broken lines indicate the four disulfide bridges. Alpha-helices are visible in the ranges 25 to 35, 90 to 100, and 120 to 125.

Other Approaches to the Determination of Protein Structure

None of the several other physical methods that have been used to obtain information on the secondary and tertiary structure of proteins provides as much direct information as the X-ray diffraction technique. Most of the techniques, however, are much simpler than X-ray diffraction, which requires, for the resolution of the structure of one protein, many years of work and equipment such as electronic computers. Some of the simpler techniques are based on the optical properties of proteins—refractivity, absorption of light of different wavelengths, rotation of the plane polarized light at different wavelengths, and luminescence.

Spectrophotometric Behaviour

Spectrophotometry of protein solutions (the measurement of the degree of absorbance of light by a protein within a specified wavelength) is useful within the range of visible light only with proteins that contain coloured prosthetic groups (the nonprotein components). Examples of such proteins include the red heme proteins of the blood, the purple pigments of the retina of the eye, green and yellow proteins that contain bile pigments, blue copper-containing proteins, and dark brown proteins called melanins. Peptide bonds, because of their carbonyl groups, absorb light energy at very short wavelengths (185–200 nanometres). The aromatic rings of phenylalanine, tyrosine, and tryptophan, however, absorb ultraviolet light between wavelengths of 280 and 290 nanometres. The absorbance of ultraviolet light by tryptophan is greatest, that of tyrosine is less, and that of phenylalanine is least. If the tyrosine or tryptophan content of the protein is known, therefore, the concentration of the protein solution can be determined by measuring its absorbance between 280 and 290 nanometres.

Optical Activity

It will be recalled that the amino acids, with the exception of glycine, exhibit optical activity (rotation of the plane of polarized light;). It is not surprising, therefore, that proteins also are optically active. They are usually levorotatory (i.e., they rotate the plane of polarization to the left) when polarized light of wavelengths in the visible range is used. Although the specific rotation (a function of the concentration of a protein solution and the distance the light travels in it) of most L-amino acids varies from $-30°$ to $+30°$, the amino acid cystine has a specific rotation of approximately $-300°$. Although the optical rotation of a protein depends on all of the amino acids of which it is composed, the most important ones are cystine and the aromatic amino acids phenylalanine, tyrosine, and tryptophan. The contribution of the other amino acids to the optical activity of a protein is negligibly small.

Chemical Reactivity of Proteins

Information on the internal structure of proteins can be obtained with chemical methods that reveal whether certain groups are present on the surface of the protein molecule and thus able to react or whether they are buried inside the closely folded peptide chains and thus are unable to react. The chemical reagents used in such investigations must be mild ones that do not affect the structure of the protein.

The reactivity of tyrosine is of special interest. It has been found, for example, that only three of the six tyrosines found in the naturally occurring enzyme ribonuclease can be iodinated (i.e., reacted to accept an iodine atom). Enzyme-catalyzed breakdown of iodinated ribonuclease is used to identify the peptides in which the iodinated tyrosines are present. The three tyrosines that can be iodinated lie on the surface of ribonuclease; the others, assumed to be inaccessible, are said to be buried in the molecule. Tyrosine can also be identified by using other techniques—e.g., treatment with diazonium compounds or tetranitromethane. Because the compounds formed are coloured, they can easily be detected when the protein is broken down with enzymes.

Cysteine can be detected by coupling with compounds such as iodoacetic acid or iodoacetamide; the reaction results in the formation of carboxymethylcysteine or carbamidomethylcysteine, which can be detected by amino acid determination of the peptides containing them. The imidazole groups of certain histidines can also be located by coupling with the same reagents under different conditions. Unfortunately, few other amino acids can be labelled without changes in the secondary and tertiary structure of the protein.

Association of Protein Subunits

Many proteins with molecular weights of more than 50,000 occur in aqueous solutions as complexes: dimers, tetramers, and higher polymers—i.e., as chains of two, four, or more repeating basic structural units. The subunits, which are called monomers or protomers, usually are present as an even number. Less than 10 percent of the polymers have been found to have an odd number of monomers. The arrangement of the subunits is thought to be regular and may be cyclic, cubic, or tetrahedral. Some of the small proteins also contain subunits. Insulin, for example, with a molecular weight of about 6,000, consists of two peptide chains linked to each other by disulfide bridges (−S−S−). Similar interchain disulfide bonds have been found in the immunoglobulins. In other proteins, hydrogen bonds and hydrophobic bonds (resulting from the interaction between the amino acid side chains of valine, leucine, isoleucine, and phenylalanine) cause the formation of aggregates of the subunits. The subunits of some proteins are identical; those of others differ. Hemoglobin is a tetramer consisting of two α-chains and two β-chains.

Protein Denaturation

When a solution of a protein is boiled, the protein frequently becomes insoluble—i.e., it is denatured—and remains insoluble even when the solution is cooled. The denaturation of the proteins of egg white by heat—as when boiling an egg—is an example of irreversible denaturation. The denatured protein has the same primary structure as the original, or native, protein. The weak forces between charged groups and the weaker forces of mutual attraction of nonpolar groups are disrupted at elevated temperatures, however; as a result, the tertiary structure of the protein is lost. In some instances the original structure of the protein can be regenerated; the process is called renaturation.

Denaturation can be brought about in various ways. Proteins are denatured by treatment with alkaline or acid, oxidizing or reducing agents, and certain organic solvents. Interesting among denaturing agents are those that affect the secondary and tertiary structure without affecting the primary structure. The agents most frequently used for this purpose are urea and guanidinium chloride. These molecules, because of their high affinity for peptide bonds, break the hydrogen bonds and the salt bridges between positive and negative side chains, thereby abolishing the tertiary structure of the peptide chain. When denaturing agents are removed from a protein solution, the native protein re-forms in many cases. Denaturation can also be accomplished by reduction of the disulfide bonds of cystine—i.e., conversion of the disulfide bond ($-S-S-$) to two sulfhydryl groups ($-SH$). This, of course, results in the formation of two cysteines. Reoxidation of the cysteines by exposure to air sometimes regenerates the native protein. In other cases, however, the wrong cysteines become bound to each other, resulting in a different protein. Finally, denaturation can also be accomplished by exposing proteins to organic solvents such as ethanol or acetone. It is believed that the organic solvents interfere with the mutual attraction of nonpolar groups.

Some of the smaller proteins, however, are extremely stable, even against heat; for example, solutions of ribonuclease can be exposed for short periods of time to temperatures of 90 °C (194 °F) without undergoing significant denaturation. Denaturation does not involve identical changes in protein molecules. A common property of denatured proteins, however, is the loss of biological activity—e.g., the ability to act as enzymes or hormones.

Although denaturation had long been considered an all-or-none reaction, it is now thought that many intermediary states exist between native and denatured protein. In some instances, however, the breaking of a key bond could be followed by the complete breakdown of the conformation of the native protein.

Although many native proteins are resistant to the action of the enzyme trypsin, which breaks down proteins during digestion, they are hydrolyzed by the same enzyme after denaturation. The peptide bonds that can be split by trypsin are inaccessible in the native proteins but become accessible during denaturation. Similarly, denatured proteins give more intense colour reactions for tyrosine, histidine, and arginine than do the same proteins in the native state. The increased accessibility of reactive groups of denatured proteins is attributed to an unfolding of the peptide chains.

If denaturation can be brought about easily and if renaturation is difficult, how is the native conformation of globular proteins maintained in living organisms, in which they are produced stepwise, by incorporation of one amino acid at a time? Experiments on the biosynthesis of proteins from amino acids containing radioactive carbon or heavy hydrogen reveal that the protein molecule grows stepwise from the N terminus to the C terminus; in each step a single amino acid residue is incorporated. As soon as the growing peptide chain contains six or seven amino acid residues, the side chains interact with each other and thus cause deviations from the straight or β-chain configuration. Depending on the nature of the side chains, this may result in the formation of an α-helix or of loops closed by hydrogen bonds or disulfide bridges. The final conformation is probably frozen when the peptide chain attains a length of 50 or more amino acid residues.

Conformation of Proteins in Interfaces

Like many other substances with both hydrophilic and hydrophobic groups, soluble proteins tend to migrate into the interface between air and water or oil and water; the term oil here means a hydrophobic liquid such as benzene or xylene. Within the interface, proteins spread, forming thin films. Measurements of the surface tension, or interfacial tension, of such films indicate that tension is reduced by the protein film. Proteins, when forming an interfacial film, are present as a monomolecular layer—i.e., a layer one molecule in height. Although it was once thought that globular protein molecules unfold completely in the interface, it has now been established that many proteins can be recovered from films in the native state. The application of lateral pressure on a protein film causes it to increase in thickness and finally to form a layer with a height corresponding to the diameter of the native protein molecule. Protein molecules in an interface, because of Brownian motions (molecular vibrations), occupy much more space than do those in the film after the application of pressure. The Brownian motion of compressed molecules is limited to the two dimensions of the interface, since the protein molecules cannot move upward or downward.

The motion of protein molecules at the air–water interface has been used to determine the molecular weight of proteins. The technique involves measuring the force exerted by the protein layer on a barrier.

When a protein solution is vigorously shaken in air, it forms a foam, because the soluble proteins migrate into the air–water interface and persist there, preventing or slowing the reconversion of the foam into a homogeneous solution. Some of the unstable, easily modified proteins are denatured when spread in the air–water interface. The formation of a permanent foam when egg white is vigorously stirred is an example of irreversible denaturation by spreading in a surface.

Classification of Proteins

Classification by Solubility

After two German chemists, Emil Fischer and Franz Hofmeister, independently stated in 1902 that proteins are essentially polypeptides consisting of many amino acids, an attempt was made to classify proteins according to their chemical and physical properties, because the biological function of proteins had not yet been established. (The protein character of enzymes was not proved until the 1920s.) Proteins were classified primarily according to their solubility in a number of solvents. This classification is no longer satisfactory, however, because proteins of quite different structure and function sometimes have similar solubilities; conversely, proteins of the same function and similar structure sometimes have different solubilities. The terms associated with the old classification, however, are still widely used. They are defined below.

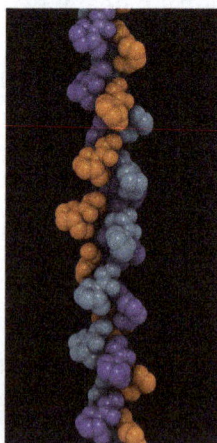

Collagen molecule.

Albumins are proteins that are soluble in water and in water half-saturated with ammonium sulfate. On the other hand, globulins are salted out (i.e., precipitated) by half-saturation with ammonium sulfate. Globulins that are soluble in salt-free water are called pseudoglobulins; those insoluble in salt-free water are euglobulins. Both prolamins and glutelins, which are plant proteins, are insoluble in water; the prolamins dissolve in 50 to 80 percent ethanol, the glutelins in acidified or alkaline solution. The term protamine is used for a number of proteins in fish sperm that consist of approximately 80 percent arginine and therefore are strongly alkaline. Histones, which are less alkaline, apparently occur only in cell nuclei, where they are bound to nucleic acids. The term scleroproteins has been used for the insoluble proteins of animal organs. They include keratin, the insoluble protein of certain epithelial tissues such as the skin or hair, and collagen, the protein of the connective tissue. A large group of proteins has been called conjugated proteins, because they are complex molecules of protein consisting of protein and nonprotein moieties. The nonprotein portion is called the prosthetic group. Conjugated proteins can be subdivided into mucoproteins, which, in addition to protein, contain carbohydrate; lipoproteins, which contain lipids; phosphoproteins, which are rich in phosphate; chromoproteins, which contain pigments such as iron-porphyrins, carotenoids, bile pigments, and melanin; and finally, nucleoproteins, which contain nucleic acid.

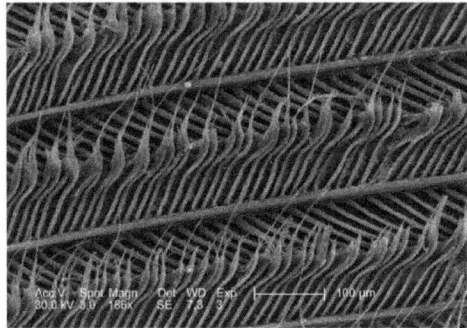

keratin: Scanning electron micrograph showing strands of keratin in a feather.

The weakness of the above classification lies in the fact that many, if not all, globulins contain small amounts of carbohydrate; thus there is no sharp borderline between globulins and mucoproteins. Moreover, the phosphoproteins do not have a prosthetic group that can be isolated; they are merely proteins in which some of the hydroxyl groups of serine are phosphorylated (i.e., contain phosphate). Finally, the globulins include proteins with quite different roles—enzymes, antibodies, fibrous proteins, and contractile proteins.

Classification by Biological Functions

In view of the unsatisfactory state of the old classification, it is preferable to classify the proteins according to their biological function. Such a classification is far from ideal, however, because one protein can have more than one function. The contractile protein myosin, for example, also acts as an ATPase (adenosine triphosphatase), an enzyme that hydrolyzes adenosine triphosphate (removes a phosphate group from ATP by introducing a water molecule). Another problem with functional classification is that the definite function of a protein frequently is not known. A protein cannot be called an enzyme as long as its substrate (the specific compound upon which it acts) is not known. It cannot even be tested for its enzymatic action when its substrate is not known.

Special Structure and Function of Proteins

Despite its weaknesses, a functional classification is used here in order to demonstrate, whenever possible, the correlation between the structure and function of a protein. The structural, fibrous proteins are presented first, because their structure is simpler than that of the globular proteins and more clearly related to their function, which is the maintenance of either a rigid or a flexible structure.

Structural Proteins

Collagen

Collagen is the structural protein of bones, tendons, ligaments, and skin. For many years collagen was considered to be insoluble in water. Part of the collagen of calf skin, however, can be extracted with citrate buffer at pH 3.7. A precursor of collagen called procollagen is converted in the body into collagen. Procollagen has a molecular weight of 120,000. Cleavage of one or a few peptide bonds of procollagen yields collagen, which has three subunits, each with a molecular weight of 95,000; therefore, the molecular weight of collagen is 285,000 (3 × 95,000). The three subunits are wound as spirals around an elongated straight axis. The length of each subunit is 2,900 angstroms, and its diameter is approximately 15 angstroms. The three chains are staggered, so that the trimer has no definite terminal limits.

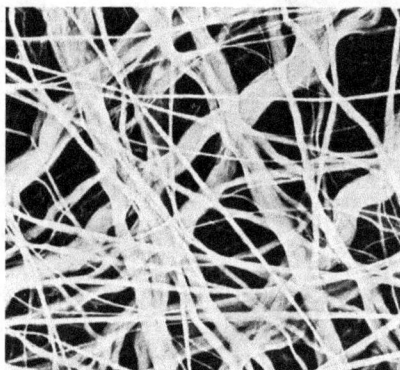

Randomly oriented collagenous fibres of varying size in a thin spread of loose areolar connective tissue (magnified about 370 ×).

Collagen differs from all other proteins in its high content of proline and hydroxyproline. Hydroxyproline does not occur in significant amounts in any other protein except elastin. Most of the proline in collagen is present in the sequence glycine–proline-X, in which X is frequently alanine or hydroxyproline. Collagen does not contain cystine or tryptophan and therefore cannot substitute for other proteins in the diet. The presence of proline causes kinks in the peptide chain and thus reduces the length of the amino acid unit from 3.7 angstroms in the extended chain of the β-structure to 2.86 angstroms in the collagen chain. In the intertwined triple helix, the glycines are inside, close to the axis; the prolines are outside.

Native collagen resists the action of trypsin but is hydrolyzed by the bacterial enzyme collagenase. When collagen is boiled with water, the triple helix is destroyed, and the subunits are partially hydrolyzed; the product is gelatin. The unfolded peptide chains of gelatin trap large amounts of water, resulting in a hydrated molecule.

When collagen is treated with tannic acid or with chromium salts, cross links form between the collagen fibres, and it becomes insoluble; the conversion of hide into leather is based on this tanning process. The tanned material is insoluble in hot water and cannot be converted to gelatin. On exposure to water at 62° to 63 °C (144° to 145 °F), however, the cross links formed by the tanning agents collapse, and the leather contracts irreversibly to about one-third its original volume.

Collagen seems to undergo an aging process in living organisms that may be caused by the formation of cross links between collagen fibres. They are formed by the conversion of some lysine side chains to aldehydes (compounds with the general structure RCHO), and the combination of the aldehydes with the ε-amino groups of intact lysine side chains. The protein elastin, which occurs in the elastic fibres of connective tissue, contains similar cross links and may result from the combination of collagen fibres with other proteins. When cross-linked collagen or elastin is degraded, products of the cross-linked lysine fragments, called desmosins and isodesmosins, are formed.

Keratin

Keratin, the structural protein of epithelial cells in the outermost layers of the skin, has been isolated from hair, nails, hoofs, and feathers. Keratin is completely insoluble in cold or hot water; it is not attacked by proteolytic enzymes (i.e., enzymes that break apart, or lyse, protein molecules), and therefore cannot replace proteins in the diet. The great stability of keratin results from the numerous disulfide bonds of cystine. The amino acid composition of keratin differs from that of collagen. Cystine may account for 24 percent of the total amino acids. The peptide chains of keratin are arranged in approximately equal amounts of antiparallel and parallel pleated sheets, in which the peptide chains are linked to each other by hydrogen bonds between the carbonyl and imino groups.

Reduction of the disulfide bonds to sulfhydryl groups results in dissociation of the peptide chains, the molecular weight of which is 25,000 to 28,000 each. The formation of permanent waves in the beauty treatment of hair is based on partial reduction of the disulfide bonds of hair keratin by thioglycol, or some other mild reducing agent, and subsequent oxidation of the sulfhydryl groups ($-SH$) in the reoriented hair to disulfide bonds ($-S-S-$) by exposure to the oxygen of the air.

The length of keratin fibres depends on their water content. They can bind approximately 16 percent of water; this hydration is accompanied by an increase in the length of the fibres of 10 to 12 percent.

The most thoroughly investigated keratin is hair keratin, particularly that of wool. It consists of a mixture of peptides with high and low cystine content. When wool is heated in water to about 90 °C (190 °F), it shrinks irreversibly. This is attributed to the breakage of hydrogen bonds and other noncovalent bonds; disulfide bonds do not seem to be affected.

The most thoroughly investigated scleroprotein has been fibroin, the insoluble material of silk. The raw silk comprising the cocoon of the silkworm consists of two proteins. One, sericin, is soluble in hot water; the other, fibroin, is not. The amino acid composition of the latter differs from that of all other proteins. It contains large amounts of glycine, alanine, tyrosine, and serine; small amounts of the other amino acids; and no sulfur-containing ones. The peptide chains are arranged in antiparallel β-structures. Fibroin is partly soluble in concentrated solutions of lithium thiocyanate or in mixtures of cupric salts and ethylene diamine. Such solutions contain a protein of molecular weight 170,000, which is a dimer of two subunits.

Little is known about either the scleroproteins of the marine sponges or the insoluble proteins of the cellular membranes of animal cells. Some of the membranes are soluble in detergents; others, however, are detergent-insoluble.

The Muscle Proteins

The total amount of muscle proteins in mammals, including humans, exceeds that of any other protein. About 40 percent of the body weight of a healthy human adult weighing about 70 kilograms (150 pounds) is muscle, which is composed of about 20 percent muscle protein. Thus, the human body contains about 5 to 6 kilograms (11 to 13 pounds) of muscle protein. An albumin-like fraction of these proteins, originally called myogen, contains various enzymes—phosphorylase, aldolase, glyceraldehyde phosphate dehydrogenase, and others; it does not seem to be involved in contraction. The globulin fraction contains myosin, the contractile protein, which also occurs in blood platelets, small bodies found in blood. Similar contractile substances occur in other contractile structures; for example, in the cilia or flagella (whiplike organs of locomotion) of bacteria and protozoans. In contrast to the scleroproteins, the contractile proteins are soluble in salt solutions and susceptible to enzymatic digestion.

The energy required for muscle contraction is provided by the oxidation of carbohydrates or lipids. The term mechanochemical reaction has been used for this conversion of chemical into mechanical energy. The molecular process underlying the reaction is known to involve the fibrous muscle proteins, the peptide chains of which undergo a change in conformation during contraction.

Myosin, which can be removed from fresh muscle by adding it to a chilled solution of dilute potassium chloride and sodium bicarbonate, is insoluble in water. Myosin, solutions of which are highly viscous, consists of an elongated—probably double-stranded—peptide chain, which is coiled at both ends in such a way that a terminal globule is formed. The length of the molecule is approximately 160 nanometres and its average diameter 2.6 nanometres. The equivalent weight of each of the two terminal globules is approximately 30,000; the molecular weight of myosin is close to 500,000. Trypsin splits myosin into large fragments called meromyosin. Myosin contains many amino acids with positively and negatively charged side chains; they form 18 and 16 percent, respectively, of the total number of amino acids. Myosin catalyzes the hydrolytic cleavage of ATP (adenosine triphosphate). A smaller protein with properties similar to those of myosin is tropomyosin. It has a molecular weight of 70,000 and dimensions of 45 by 2 nanometres. More than 90 percent of its peptide chains are present in the α-helix form.

Myosin combines easily with another muscle protein called actin, the molecular weight of which is about 50,000; it forms 12 to 15 percent of the muscle proteins. Actin can exist in two forms—one, G-actin, is globular; the other, F-actin, is fibrous. Actomyosin is a complex molecule formed by one molecule of myosin and one or two molecules of actin. In muscle, actin and myosin filaments are oriented parallel to each other and to the long axis of the muscle. The actin filaments are linked to each other lengthwise by fine threads called S filaments. During contraction the S filaments shorten, so that the actin filaments slide toward each other, past the myosin filaments, thus causing a shortening of the muscle.

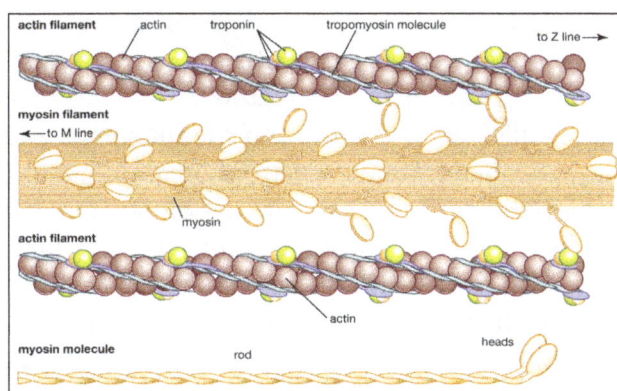

Muscle: actin and myosin
The structure of actin and myosin filaments.

Fibrinogen and Fibrin

Fibrinogen, the protein of the blood plasma, is converted into the insoluble protein fibrin during the clotting process. The fibrinogen-free fluid obtained after removal of the clot, called blood serum, is blood plasma minus fibrinogen. The fibrinogen content of the blood plasma is 0.2 to 0.4 percent.

Fibrinogen can be precipitated from the blood plasma by half-saturation with sodium chloride. Fibrinogen solutions are highly viscous and show strong flow birefringence. In electron micrographs the molecules appear as rods with a length of 47.5 nanometres and a diameter of 1.5 nanometres; in addition, two terminal and a central nodule are visible. The molecular weight is 340,000. An unusually high percentage, about 36 percent, of the amino acid side chains are positively or negatively charged.

The clotting process is initiated by the enzyme thrombin, which catalyzes the breakage of a few peptide bonds of fibrinogen; as a result, two small fibrinopeptides with molecular weights of 1,900 and 2,400 are released. The remainder of the fibrinogen molecule, a monomer, is soluble and stable at pH values less than 6 (i.e., in acid solutions). In neutral solution (pH 7) the monomer is converted into a larger molecule, insoluble fibrin; this results from the formation of new peptide bonds. The newly formed peptide bonds form intermolecular and intramolecular cross links, thus giving rise to a large clot, in which all molecules are linked to each other. Clotting, which takes place only in the presence of calcium ions, can be prevented by compounds such as oxalate or citrate, which have a high affinity for calcium ions.

Albumins, Globulins and other Soluble Proteins

The blood plasma, the lymph, and other animal fluids usually contain one to seven grams of protein per 100 millilitres of fluid, which includes small amounts of hundreds of enzymes and a large number of protein hormones. The discussion below is limited largely to the proteins that occur in large amounts and can be easily isolated from the body fluids.

Proteins of the Blood Serum

Human blood serum contains about 7 percent protein, two-thirds of which is in the albumin fraction; the other third is in the globulin fraction. Electrophoresis of serum reveals a large albumin peak and three smaller globulin peaks, the alpha-, beta-, and gamma-globulins. The amounts of alpha-, beta-, and gamma-globulin in normal human serum are approximately 1.5, 1.9, and 1.1 percent, respectively. Each globulin fraction is a mixture of many different proteins, as has been demonstrated by immunoelectrophoresis. In this method, serum from an animal (e.g., a rabbit) injected with human serum is allowed to diffuse into the four protein bands—albumin, alpha-, beta-, and gamma-globulin—obtained from the electrophoresis of human serum. Because the animal has previously been injected with human serum, its blood contains antibodies (substances formed in response to a foreign substance introduced into the body) against each of the human serum proteins; each antibody combines with the serum protein (antigen) that caused its formation in the animal. The result is the formation of about 20 regions of insoluble antigen-antibody precipitate, which appear as white arcs in the transparent gel of the electrophoresis medium. Each region corresponds to a different human serum protein.

Serum albumin is much less heterogeneous (i.e., contains fewer distinct proteins) than are the globulins; in fact, it is one of the few serum proteins that can be obtained in a crystalline form. Serum albumin combines easily with many acidic dyes (e.g., Congo red and methyl orange); with bilirubin, the yellow bile pigment; and with fatty acids. It seems to act, in living organisms, as a carrier for certain biological substances. Present in blood serum in relatively high concentration, serum albumin also acts as a protective colloid, a protein that stabilizes other proteins. Albumin (molecular weight of 68,000) has a single free sulfhydryl ($-SH$) group, which on oxidation forms a disulfide bond with the sulfhydryl group of another serum albumin molecule, thus forming a dimer. The isoelectric point of serum albumin is pH 4.7.

The alpha-globulin fraction of blood serum is a mixture of several conjugated proteins. The best known are an α-lipoprotein (combination of lipid and protein) and two mucoproteins (combinations of carbohydrate and protein). One mucoprotein is called orosomucoid, or α1-acid glycoprotein; the other is called haptoglobin because it combines specifically with globin, the protein component of hemoglobin. Haptoglobin contains about 20 percent carbohydrate. The beta-globulin fraction of serum contains, in addition to lipoproteins and mucoproteins, two metal-binding proteins, transferrin and ceruloplasmin, which bind iron and copper, respectively. They are the principal iron and copper carriers of the blood.

The gamma-globulins are the most heterogeneous globulins. Although most have a molecular weight of approximately 150,000, that of some, called macroglobulins, is as high as 800,000. Because typical antibodies are of the same size and exhibit the same electrophoretic behaviour as γ-globulins, they are called immunoglobulins. The designation IgM or gamma M (γM) is used for the macroglobulins; the designation IgG or gamma G (γG) is used for γ–globulins of molecular weight 150,000.

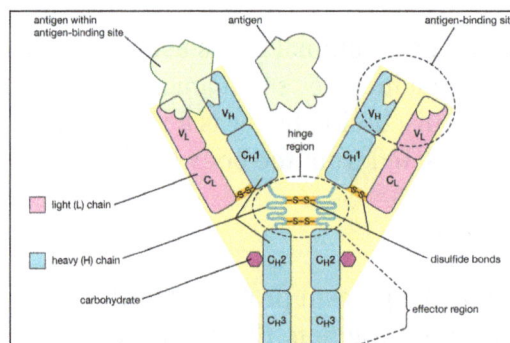

The four-chain structure of an antibody, or immunoglobulin, molecule.

The basic unit is composed of two identical light (L) chains and two identical heavy (H) chains, which are held together by disulfide bonds to form a flexible Y shape. Each chain is composed of a variable (V) region and a constant (C) region.

Milk Proteins

Milk contains the following: an albumin, α-lactalbumin; a globulin, beta-lactoglobulin; and a phosphoprotein, casein. If acid is added to milk, casein precipitates. The remaining watery liquid (the supernatant solution), or whey, contains α-lactalbumin and β-lactoglobulin. Both have been obtained in crystalline form; in bovine milk, their molecular weights are approximately 14,000 and 18,400, respectively. Lactoglobulin also occurs as a dimer of molecular weight 37,000. Genetic variations can produce small variations in the amino acid composition of lactoglobulin. The amino acid composition and the tertiary structure of lactalbumin resemble that of lysozyme, an egg protein.

Casein is precipitated not only by the addition of acid but also by the action of the enzyme rennin, which is found in gastric juice. Rennin from calf stomachs is used to precipitate casein, from which cheese is made. Milk fat precipitates with casein; milk sugar, however, remains in the supernatant (whey). Casein is a mixture of several similar phosphoproteins, called α-, β-, γ-, and κ-casein, all of which contain some serine side chains combined with phosphoric acid. Approximately 75 percent of casein is α-casein. Cystine has been found only in κ-casein. In milk, casein seems to form polymeric globules (micelles) with radially arranged monomers, each with a molecular weight of 24,000; the acidic side chains occur predominantly on the surface of the micelle, rather than inside.

Egg Proteins

About 50 percent of the proteins of egg white are composed of ovalbumin, which is easily obtained in crystals. Its molecular weight is 46,000 and its amino acid composition differs from that of serum albumin. Other proteins of egg white are conalbumin, lysozyme, ovoglobulin, ovomucoid, and avidin. Lysozyme is an enzyme that hydrolyzes the carbohydrates found in the capsules certain bacteria secrete around themselves; it causes lysis (disintegration) of the bacteria. The molecular weight of lysozyme is 14,100. Its three-dimensional structure is similar to that of α-lactalbumin, which stimulates the formation of lactose by the enzyme lactose synthetase. Lysozyme has also been found in the urine of patients suffering from leukemia, meningitis, and renal disease.

Avidin is a glycoprotein that combines specifically with biotin, a vitamin. In animals fed large amounts of raw egg white, the action of avidin results in "egg-white injury." The molecular weight of avidin, which forms a tetramer, is 16,200. Its amino acid sequence is known.

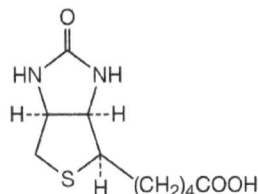

Egg-yolk proteins contain a mixture of lipoproteins and livetins. The latter are similar to serum albumin, α-globulin, and β-globulin. The yolk also contains a phosphoprotein, phosvitin. Phosvitin, which has also been found in fish sperm, has a molecular weight of 40,000 and an unusual amino acid composition; one third of its amino acids are phosphoserine.

Protamines and Histones

Protamines are found in the sperm cells of fish. The most thoroughly investigated protamines are salmine from salmon sperm and clupeine from herring sperm. The protamines are bound to deoxyribonucleic acid (DNA), forming nucleoprotamines. The amino acid composition of the protamines is simple; they contain, in addition to large amounts of arginine, small amounts of five or six other amino acids. The composition of the salmine molecule, for example, is: Arg51, Ala4, Val4, Ile1, Pro7, and Ser6, in which the subscript numbers indicate the number of each amino acid in the molecule. Because of the high arginine content, the isoelectric points of the protamines are at pH values of 11 to 12; i.e., the protamines are alkaline. The molecular weights of salmine and clupeine are close to 6,000. All of the protamines investigated thus far are mixtures of several similar proteins.

The histones are less basic than the protamines. They contain high amounts of either lysine or arginine and small amounts of aspartic acid and glutamic acid. Histones occur in combination with DNA as nucleohistones in the nuclei of the body cells of animals and plants, but not in animal sperm. The molecular weights of histones vary from 10,000 to 22,000. In contrast to the protamines, the histones contain most of the 20 amino acids, with the exception of tryptophan and the sulfur-containing ones. Like the protamines, histone preparations are heterogeneous mixtures. The amino acid sequence of some of the histones has been determined.

Plant Proteins

Plant proteins, mostly globulins, have been obtained chiefly from the protein-rich seeds of cereals and legumes. Small amounts of albumins are found in seeds. The best known globulins, insoluble in water, can be extracted from seeds by treatment with 2 to 10 percent solutions of sodium chloride. Many plant globulins have been obtained in crystalline form; they include edestin from hemp, molecular weight 310,000; amandin from almonds, 330,000; concanavalin A (42,000) and B (96,000); and canavalin (113,000) from jack beans. They are polymers of smaller subunits; edestin, for example, is a hexamer of a subunit with a molecular weight of 50,000, and concanavalin B a trimer of a subunit with a molecular weight of 30,000. After extraction of lipids from cereal seeds by ether and alcohol, further extraction with water containing 50 to 80 percent of alcohol yields proteins that are insoluble in water but soluble in water–ethanol mixtures and have been called prolamins. Their solubility in aqueous ethanol may result from their high proline and glutamine content. Gliadin, the prolamin from wheat, contains 14 grams of proline and 46 grams of glutamic acid in 100 grams of protein; most of the glutamic acid is in the form of glutamine. The total amounts of the basic amino acids (arginine, lysine, and histidine) in gliadin are only 5 percent of the weight of gliadin. Because the glysine content is either low or nonexistent, human populations dependent on grain as a sole protein source suffer from lysine deficiency.

Conjugated Proteins

Combination of Proteins with Prosthetic Groups

The link between a protein molecule and its prosthetic group is a covalent bond (an electron-sharing bond) in the glycoproteins, the biliproteins, and some of the heme proteins. In lipoproteins, nucleoproteins, and some heme proteins, the two components are linked by noncovalent bonds; the bonding results from the same forces that are responsible for the tertiary structure of proteins: Hydrogen bonds, salt bridges between positively and negatively charged groups, disulfide bonds, and mutual interaction of hydrophobic groups. In the metalloproteins (proteins with a metal element as a prosthetic group), the metal ion usually forms a centre to which various groups are bound.

Some of the conjugated proteins have been mentioned in preceding sections because they occur in the blood serum, in milk, and in eggs; others are discussed below in sections dealing with respiratory proteins and enzymes.

Mucoproteins and Glycoproteins

The prosthetic groups in mucoproteins and glycoproteins are oligosaccharides (carbohydrates consisting of a small number of simple sugar molecules) usually containing from four to 12 sugar molecules; the most common sugars are galactose, mannose, glucosamine, and galactosamine. Xylose, fucose, glucuronic acid, sialic acid, and other simple sugars sometimes also occur. Some mucoproteins contain 20 percent or more of carbohydrate, usually in several oligosaccharides attached to different parts of the peptide chain. The designation mucoprotein is used for proteins with more than 3 to 4 percent carbohydrate; if the carbohydrate content is less than 3 percent, the protein is sometimes called a glycoprotein or simply a protein.

Mucoproteins, highly viscous proteins originally called mucins, are found in saliva, in gastric juice, and in other animal secretions. Mucoproteins occur in large amounts in cartilage, synovial fluid (the lubricating fluid of joints and tendons), and egg white. The mucoprotein of cartilage is formed by the combination of collagen with chondroitinsulfuric acid, which is a polymer of either glucuronic or iduronic acid and acetylhexosamine or acetylgalactosamine. It is not yet clear whether or not chondroitinsulfate is bound to collagen by covalent bonds.

Lipoproteins and Proteolipids

The bond between the protein and the lipid portion of lipoproteins and proteolipids is a noncovalent one. It is thought that some of the lipid is enclosed in a meshlike arrangement of peptide chains and becomes accessible for reaction only after the unfolding of the chains by denaturing agents. Although lipoproteins in the α- and β-globulin fraction of blood serum are soluble in water (but insoluble in organic solvents), some of the brain lipoproteins, because they have a high lipid content, are soluble in organic solvents; they are called proteolipids. The β-lipoprotein of human blood serum is a macroglobulin with a molecular weight of about 1,300,000, 70 percent of which is lipid; of the lipid, about 30 percent is phospholipid and 40 percent cholesterol and compounds derived from it. Because of their lipid content, the lipoproteins have the lowest density (mass per unit volume) of all proteins and are usually classified as low- and high-density lipoproteins (LDL and HDL).

Coloured lipoproteins are formed by the combination of protein with carotenoids. Crustacyanin, the pigment of lobsters, crayfish, and other crustaceans, contains astaxanthin, which is a compoundderived from carotene. Among the most interesting of the coloured lipoproteins are the pigments of the retina of the eye. They contain retinal, which is a compound derived from carotene and which is formed by the oxidation of vitamin A. In rhodopsin, the red pigment of the retina, the aldehyde group ($-CHO$) of retinal forms a covalent bond with an amino ($-NH_2$) group of opsin, the protein carrier. Colour vision is mediated by the presence of several visual pigments in the retina that differ from rhodopsin either in the structure of retinal or in that of the protein carrier.

Metalloproteins

Proteins in which heavy metal ions are bound directly to some of the side chains of histidine, cysteine, or some other amino acid are called metalloproteins. Two metalloproteins, transferrin and ceruloplasmin, occur in the globulin fractions of blood serum; they act as carriers of iron and copper, respectively. Transferrin has a molecular weight of about 80,000 and consists of two identical subunits, each of which contains one ferric ion (Fe^{3+}) that seems to be bound to tyrosine. Several genetic variants of transferrin are known to occur in humans. Another iron protein, ferritin, which contains 20 to 22 percent iron, is the form in which iron is stored in animals; it has been obtained in crystalline form from liver and spleen. A molecule consisting of 20 subunits, its molecular weight is approximately 480,000. The iron can be removed by reduction from the ferric (Fe^{3+}) to the ferrous (Fe^{2+}) state. The iron-free protein, apoferritin, is synthesized in the body before the iron is incorporated.

Green plants and some photosynthetic and nitrogen-fixing bacteria (i.e., bacteria that convert atmospheric nitrogen, N_2, into amino acids and proteins) contain various ferredoxins. They are small proteins containing 50 to 100 amino acids and a chain of iron and disulfide units (FeS_2), in which some of the sulfur atoms are contributed by cysteine; others are sulfide ions (S^{2-}). The number of FeS_2 units per ferredoxin molecule varies from five in the ferredoxin of spinach to 10 in the ferredoxin of certain bacteria. Ferredoxins act as electron carriers in photosynthesis and in nitrogen fixation.

Ceruloplasmin is a copper-containing globulin that has a molecular weight of 151,000; the molecule consists of eight subunits, each containing one copper ion. Ceruloplasmin is the principal carrier of copper in organisms, although copper can also be transported by the iron-containing globulin transferrin. Another copper-containing protein, copper-zinc superoxide dismutase (formerly known as erythrocuprein), has been isolated from red blood cells; it has also been found in the liver and in the brain. The molecule, which consists of two subunits of similar size, contains copper ions and zinc ions. Because of their copper content, ceruloplasmin and copper-zinc superoxide dismutase possess catalytic activity in oxidation-reduction reactions.

Many animal enzymes contain zinc ions, which are usually bound to the sulfur of cysteine. Horse kidneys contain the protein metallothionein, which contain zinc and cadmium; both are bound to sulfur. A vanadium-protein complex (hemovanadin) has been found in surprisingly high amounts in yellowish-green cells (vanadocytes) of tunicates, which are marine invertebrates.

Heme Proteins and other Chromoproteins

Although the heme proteins contain iron, they are usually not classified as metalloproteins, because their prosthetic group is an iron-porphyrin complex in which the iron is bound very firmly. The intense red or brown colour of the heme proteins is not caused by iron but by porphyrin, a complex cyclic structure. All porphyrin compounds absorb light intensely at or close to 410 nanometres. Porphyrin consists of four pyrrole rings (five-membered closed structures containing one nitrogen and four carbon atoms) linked to each other by methine groups ($-CH=$). The iron atom is kept in the centre of the porphyrin ring by interaction with the four nitrogen atoms. The iron atom can combine with two other substituents; in oxyhemoglobin, one substituent is a histidine of the protein carrier, the other is an oxygen molecule. In some heme proteins, the protein is also bound covalently to the side chains of porphyrin.

The chromoprotein melanin, a pigment found in dark skin, dark hair, and melanotic tumours, occurs in every major group of living organisms and appears to be remarkably diverse in structure. In humans, melanin produced by melanocytes may be dark brown (eumelanin) or pale red or yellowish (phaeomelanin). The different types are synthesized via different pathways, though they share the same initial step—the oxidation of tyrosine.

Green chromoproteins called biliproteins are found in many insects, such as grasshoppers, and also in the eggshells of many birds. The biliproteins are derived from the bile pigment biliverdin, which in turn is formed from porphyrin; biliverdin contains four pyrrole rings and three of the four methine groups of porphyrin. Large amounts of biliproteins have been found in red algae and blue-green algae; the red protein is called phycoerythrin, the blue one phycocyanobilin.

Blue-green algae in Morning Glory Pool, Yellowstone National Park, Wyoming.

Nucleoproteins

When a protein solution is mixed with a solution of a nucleic acid, the phosphoric acid component of the nucleic acid combines with the positively charged ammonium groups ($-NH_3^+$) of the protein to form a protein–nucleic acid complex. The nucleus of a cell contains predominantly deoxyribonucleic acid (DNA) and the cytoplasm predominantly ribonucleic acid (RNA); both parts of the cell also contain protein. Protein–nucleic acid complexes, therefore, form in living cells.

The only nucleoproteins for which some evidence for specificity exists are nucleoprotamines, nucleohistones, and some RNA and DNA viruses. The nucleoprotamines are the form in which protamines occur in the sperm cells of fish; the histones of the thymus and of pea seedlings and other plant material apparently occur predominantly as nucleohistones. Both nucleoprotamines and nucleohistones contain only DNA.

Some of the simplest viruses consist of a specific RNA, which is coated by protein. One of the best known RNA viruses, tobacco mosaic virus (TMV), has the shape of a rod. RNA comprises only 5.1

percent of the mass of the virus. The complete sequence of the virus protein, which consists of about 2,130 identical peptide chains, each containing 158 amino acids, has been determined. The protein is arranged in a spiral around the RNA core.

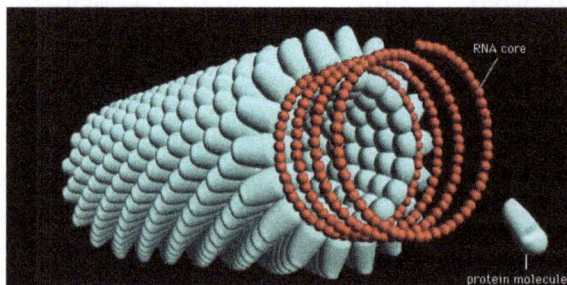

Schematic structure of the tobacco mosaic virus. The cutaway section shows the helical ribonucleic acid associated with protein molecules in a ratio of three nucleotides per protein molecule.

DNA has been found in most bacterial viruses (bacteriophages) and in some animal viruses. As in TMV, the core of DNA is surrounded by protein. Phage protein is a mixture of enzymes and therefore cannot be considered as the protein portion of only one nucleoprotein.

Respiratory Proteins

Hemoglobin

Hemoglobin is the oxygen carrier in all vertebrates and some invertebrates. In oxyhemoglobin (HbO_2), which is bright red, the ferrous ion (Fe^{2+}) is bound to the four nitrogen atoms of porphyrin; the other two substituents are an oxygen molecule and the histidine of globin, the protein component of hemoglobin. Deoxyhemoglobin (deoxy-Hb), as its name implies, is oxyhemoglobin minus oxygen (i.e., reduced hemoglobin); it is purple in colour. Oxidation of the ferrous ion of hemoglobin yields a ferric compound, methemoglobin, sometimes called hemiglobin or ferrihemoglobin. The oxygen of oxyhemoglobin can be displaced by carbon monoxide, for which hemoglobin has a much greater affinity, preventing oxygen from reaching the body tissues.

The hemoglobins of all mammals, birds, and many other vertebrates are tetramers of two α- and two β-chains. The molecular weight of the tetramer is 64,500; the molecular weight of the α- and β-chains is approximately 16,100 each, and the four subunits are linked to each other by noncovalent interactions. If hemin (the ferric porphyrin component) is removed from globin (the protein component), two molecules of globin, each consisting of one α- and one β-chain, are obtained; the molecular weight of globin is 32,200. In contrast to hemoglobin, globin is an unstable protein that is easily denatured. If native globin is incubated with a solution of hemin at pH values of 8 to 9, native hemoglobin is reconstituted. Myoglobin, the red pigment of mammalian muscles, is a monomer with a molecular weight of 16,000.

The mammalian hemoglobins differ from each other in their amino acid composition and therefore in their secondary and tertiary structure. Rat and horse hemoglobins crystallize very easily, but those of humans, cattle, and sheep, because they are more soluble, are difficult to crystallize. The shape of hemoglobin crystals varies in different species; moreover, decomposition and denaturation occur at different rates in different species. It was also found that the blood of human newborns contains two different hemoglobins: About 20 percent of their hemoglobin is an adult hemoglobin (hemoglobin A) and 80 percent is a fetal hemoglobin (hemoglobin F). Hemoglobin F persists in the infant for the first seven months of life. The same hemoglobin F has also been found in the blood of patients suffering from thalassemia, an anemia with a high incidence in regions surrounding the Mediterranean Sea. Hemoglobin F contains, as does hemoglobin A, two α-chains; the two β-chains, however, have been replaced by two quite different γ-chains. When the technique of electrophoresis was first applied to the hemoglobin of blacks suffering from sickle cell anemia in 1949, a new hemoglobin (hemoglobin S) was discovered. More than 200 different human hemoglobins have been discovered since. They differ from normal hemoglobin A in the amino acid composition of either the α- or the β-chain.

The hemoglobins of some of the lowest fishes are monomers containing one iron atom per molecule. Hemoglobin-like respiratory proteins have been found in some invertebrates. The red hemoglobin of insects, mollusks, and protozoans is called erythrocruorin. It differs from vertebrate hemoglobin by its high molecular weight.

Although green plants contain no hemoglobin, a red protein, called leghemoglobin, has been discovered in the root nodules of leguminous plants. It seems to be produced by the nitrogen-fixing bacteria of the root nodules and may be involved in the reduction of atmospheric nitrogen to ammonia and amino acids.

Other Respiratory Proteins

A green respiratory protein, chlorocruorin, has been found in the blood of marine worms in the genera Serpula and Spirographis. It has the same high molecular weight as erythrocruorin but differs from hemoglobin in its prosthetic group. A red metalloprotein, hemerythrin, acts as a respiratory protein in marine worms of the phylum Sipuncula. The molecule consists of eight subunits with a molecular weight of 13,500 each. Hemerythrin contains no porphyrins and therefore is not a heme protein.

A metalloprotein containing copper is the respiratory protein of crustaceans (shrimps, crabs, etc.) and of some gastropods (snails). The protein, called hemocyanin, is pale yellow when not combined with oxygen, and blue when combined with oxygen. The molecular weights of hemocyanins vary from 300,000 to 9,000,000. Each animal investigated thus far apparently has a species-specific hemocyanin.

Protein Hormones

Some hormones that are products of endocrine glands are proteins or peptides, others are steroids. None of the hormones has any enzymatic activity. Each has a target organ in which it elicits some biological action—e.g., secretion of gastric or pancreatic juice, production of milk, production of steroid hormones. The mechanism by which the hormones exert their effects is not fully understood. Cyclic adenosine monophosphate is involved in the transmittance of the hormonal stimulus to the cells whose activity is specifically increased by the hormone.

Hormones of the Thyroid Gland

Thyroglobulin, the active groups of which are two molecules of the iodine-containing compoundthyroxine, has a molecular weight of 670,000. Thyroglobulin also contains thyroxine with two and three iodine atoms instead of four and tyrosine with one and two iodine atoms. Injection of the hormone causes an increase in metabolism; lack of it results in a slowdown.

Another hormone, calcitonin, which lowers the calcium level of the blood, occurs in the thyroid gland. The amino acid sequences of calcitonin from pig, beef, and salmon differ from human calcitonin in some amino acids. All of them, however, have the half-cystines (C) and the prolinamide (P) in the same position.

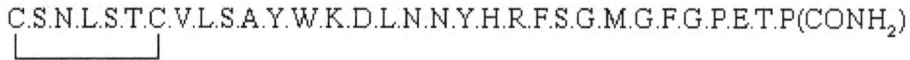

$$\text{C.S.N.L.S.T.C.V.L.S.A.Y.W.K.D.L.N.N.Y.H.R.F.S.G.M.G.F.G.P.E.T.P(CONH}_2\text{)}$$

Parathyroid hormone (parathormone), produced in small glands that are embedded in or lie behind the thyroid gland, is essential for maintaining the calcium level of the blood. A decrease in its production results in hypocalcemia (a reduction of calcium levels in the bloodstream below the normal range). Bovine parathormone has a molecular weight of 8,500; it contains no cystine or cysteine and is rich in aspartic acid, glutamic acid, or their amides.

Hormones of the Pancreas

Although the amino acid structure of insulin has been known since 1949, repeated attempts to synthesize it gave very poor yields because of the failure of the two peptide chains to combine forming the correct disulfide bridge. The ease of the biosynthesis of insulin is explained by the discovery in the pancreas of proinsulin, from which insulin is formed. The single peptide chain of proinsulin loses a peptide consisting of 33 amino acids and called the connecting peptide, or C peptide, during its conversion to insulin. The disulfide bridges of proinsulin connect the A and B chains.

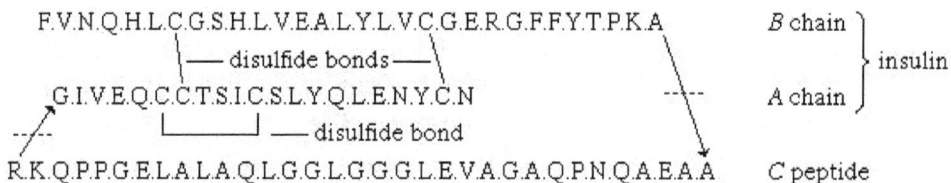

F.V.N.Q.H.L.C.G.S.H.L.V.E.A.L.Y.L.V.C.G.E.R.G.F.F.Y.T.P.K.A   B chain
       — disulfide bonds —   insulin
G.I.V.E.Q.C.C.T.S.I.C.S.L.Y.Q.L.E.N.Y.C.N   A chain
              — disulfide bond
R.K.Q.P.P.G.E.L.A.L.A.Q.L.G.G.L.G.G.G.L.E.V.A.G.A.Q.P.N.Q.A.E.A.A   C peptide

In aqueous solutions, insulin exists predominantly as a complex of six subunits, each of which contains an A and a B chain. The insulins of several species have been isolated and analyzed; their amino acid sequences have been found to differ somewhat, but all apparently contain the same disulfide bridges between the two chains.

Although the injection of insulin lowers the blood sugar, administration of glucagon, another pancreas hormone, raises the blood sugar level. Glucagon consists of a straight peptide chain of 29 amino acids. It has been synthesized; the synthetic product has the full biological activity of natural glucagon. The structure of glucagon is free of cystine and isoleucine.

The pituitary gland has an anterior lobe, a posterior lobe, and an intermediate portion; they differ in cellular structure and in the structure and action of the hormones they form. The posterior lobe produces two similar hormones, oxytocin and vasopressin. The former causes contraction of the pregnant uterus; the latter raises the blood pressure. Both are octapeptides formed by a ring of five amino acids (the two cystine halves count as one amino acid) and a side chain of three amino acids. The two cystine halves are linked to each other by a disulfide bond, and the C terminal amino acid is glycinamide. The structure has been established and confirmed. Human vasopressin differs from oxytocin in that isoleucine is replaced by phenylalanine and leucine by arginine.

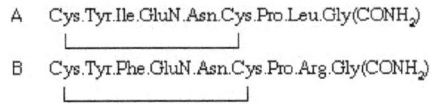

A   Cys.Tyr.Ile.GluN.Asn.Cys.Pro.Leu.Gly(CONH$_2$)

B   Cys.Tyr.Phe.GluN.Asn.Cys.Pro.Arg.Gly(CONH$_2$)

The intermediate part of the pituitary gland produces the melanocyte-stimulating hormone (MSH), which causes expansion of the pigmented melanophores (cells) in the skin of frogs and other batrachians. Two hormones, called α-MSH and β-MSH, have been prepared from hog pituitary glands. The first, α-MSH, consists of 13 amino acids; its N terminal serine is acetylated (i.e., the acetyl group, CH$_3$CO, of acetic acid is attached), and its C terminal valine residue is present as valinamide. The second, β-MSH, contains in its 18 amino acids many of those occurring in α-MSH.

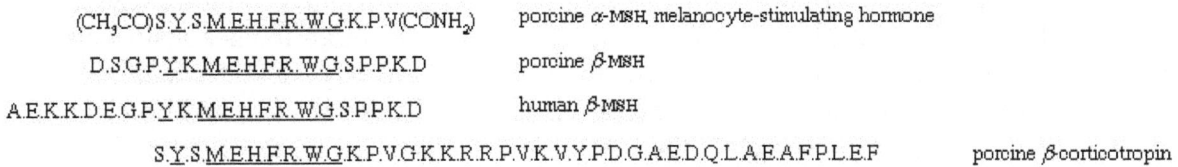

(CH$_3$CO)S.Y.S.M.E.H.F.R.W.G.K.P.V(CONH$_2$)   porcine α-MSH, melanocyte-stimulating hormone

D.S.G.P.Y.K.M.E.H.F.R.W.G.S.P.P.K.D   porcine β-MSH

A.E.K.K.D.E.G.P.Y.K.M.E.H.F.R.W.G.S.P.P.K.D   human β-MSH

S.Y.S.M.E.H.F.R.W.G.K.P.V.G.K.K.R.R.P.V.K.V.Y.P.D.G.A.E.D.Q.L.A.E.A.F.P.L.E.F   porcine β-corticotropin

The anterior pituitary lobe produces several protein hormones—a thyroid-stimulating hormone (thyrotropin), molecular weight 28,000; a lactogenic hormone, molecular weight 22,500; a growth hormone, molecular weight 21,500; a luteinizing hormone, molecular weight 30,000; and a follicle-stimulating hormone, molecular weight 29,000. The thyroid-stimulating hormone consists of α and β subunits with a composition similar to the subunits of luteinizing hormone. When separated, neither of the two subunits has hormonal activity; when combined, however, they regain about 50 percent of the original activity. The lactogenic hormone (prolactin) from sheep pituitary glands contains 190 amino acids. Their sequence has been elucidated; a similar peptide chain of 188 amino acids that has been synthesized not only has 10 percent of the biological activity of the natural hormone but also some activity of the growth hormone. The amino acid sequence of the growth hormone (somatotropic hormone) is also known; it seems to stimulate the synthesis of RNA and in this way to accelerate growth. The luteinizing hormone, a mucoprotein containing about 12 percent carbohydrate, consists of two subunits, each with a molecular weight of approximately 15,000; when separated, the subunits recombine spontaneously. The urine of pregnant women contains chorionic gonadotropin, the presence of which makes possible early diagnosis of pregnancy. The amino acid sequence is known. The sequence of 160 of its 190 amino acids is identical with those of the growth hormone; 100 of these also occur in the same sequence as in lactogenic hormone. The different pituitary hormones and the chorionic gonadotropin thus may have been derived from a common substance that, during evolution, underwent differentiation.

Peptides with Hormonelike Activity

Small peptides have been discovered that, like hormones, act on certain target organs. One peptide, angiotensin (angiotonin or hypertensin), is formed in the blood from angiotensinogen by the action of renin, an enzyme of the kidney. It is an octapeptide and increases blood pressure.

Similar peptides include bradykinin, which stimulates smooth muscles; gastrin, which stimulates secretion of hydrochloric acid and pepsin in the stomach; secretin, which stimulates the flow of pancreatic juice; and kallikrein, the activity of which is similar to bradykinin.

Immunoglobulins and Antibodies

Antibodies, proteins that combat foreign substances in the body, are associated with the globulinfraction of the immune serum. As stated previously, when the serum globulins are separated into α-, β-, and γ- fractions, antibodies are associated with the γ-globulins. Antibodies can be purified by precipitation with the antigen (i.e., the foreign substance) that caused their formation, followed by separation of the antigen-antibody complex. Antibodies prepared in this way consist of a mixture of many similar antibody molecules, which differ in molecular weight, amino acid composition, and other properties. The same differences are found in the γ-globulins of normal blood serums. The γ-globulin of normal blood serum is thought to consist of a mixture of hundreds of different γ-globulins, each of which occurs in amounts too small for isolation. Because the physical and chemical properties of normal γ-globulins are the same as those of antibodies, the γ-globulins are frequently called immunoglobulins. They may be considered to be antibodies against unknown antigens. If solutions of γ-globulin are resolved by gel filtration through dextran, the first fraction has a molecular weight of 900,000. This fraction is called IgM or γM; Ig is an abbreviation for immunoglobulin and M for macroglobulin. The next two fractions are IgA (γA) and IgG (γG), with molecular weights of about 320,000 and 150,000 respectively. Two other immunoglobulins, known as IgD and IgE, have also been detected in much smaller amounts in some immune sera.

The bulk of the immunoglobulins is found in the IgG fraction, which also contains most of the antibodies. The IgM molecules are apparently pentamers—aggregates of five of the IgG molecules. Electron microscopy shows their five subunits to be linked to each other by disulfide bonds in the form of a pentagon. The IgA molecules are found principally in milk and in secretions of the intestinal mucosa. Some of them contain, in addition to a dimer of IgG, a "secretory piece" that enables the passage of IgA molecules between tissue and fluid; the structure of the secretory piece is not yet known. The IgM and IgA immunoglobulins and antibodies contain 10 to 15 percent carbohydrate; the carbohydrate content of the IgG molecules is 2 to 3 percent.

IgG molecules treated with the enzyme papain split into three fragments of almost identical molecular weight of 50,000. Two of these, called Fab fragments, are identical; the third is abbreviated Fc. Reduction to sulfhydryl groups of some of the disulfide bonds of IgG results in the formation of two heavy, or H, chains (molecular weight 55,000) and two light, or L, chains (molecular weight 22,000). They are linked by disulfide bonds in the order $L-H-H-L$. Each H chain contains four intrachain disulfide bonds, and each L chain contains two.

Antibody preparations of the IgG type, even after removal of IgM and IgA antibodies, are heterogeneous. The H and L chains consist of a large number of different L chains and a variety of H chains. Pure IgG, IgM, and IgA immunoglobulins, however, occur in the blood serum of patients suffering from myelomas, which are malignant tumours of the bone marrow. The tumours produce either an IgG, an IgM, or an IgA protein, but rarely more than one class. A protein called the Bence-Jones protein, which is found in the urine of patients suffering from myeloma tumours, is identical with the L chains of the myeloma protein. Each patient has a different Bence-Jones protein; no two of the more than 100 Bence-Jones proteins that have been analyzed thus far are identical. It is thought that one lymphoid cell among hundreds of thousands becomes malignant and multiplies rapidly, forming the mass of a myeloma tumour that produces one γ-globulin.

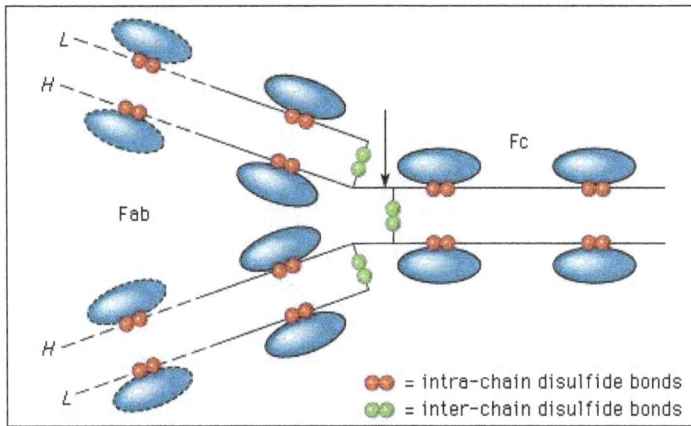

IgG immunoglobulin.

Analyses of the Bence-Jones proteins have revealed that the L chains of humans and other mammals are of two quite different types, kappa (κ) and lambda (λ). Both consist of approximately 220 amino acids. The N–terminal halves of κ- and λ-chains are variable, differing in each Bence-Jones protein. The C–terminal halves of these same L chains have a constant amino acid sequence of either the κ- or the λ-type. The fact that one half of a peptide chain is variable and the other half invariant is contradictory to the view that the amino acid sequence of each peptide chain is determined by one gene. Evidently, two genes, one of them variable, the other invariant, fuse to form the gene for the single peptide chain of the L chains. Whereas the normal human L chains are always mixtures of the κ- and λ-types, the H chains of IgG, IgM, and IgA are different. They have been designated as gamma (γ), mu (μ), and alpha (α) chains, respectively. The N-terminal quarter of the H chains has a variable amino acid sequence; the C-terminal three-quarters of the H chains have a constant amino acid sequence.

Some of the amino acid sequences in the L and H chains are transmitted from generation to generation. As a result, the constant portion of the human L chains of the κ-type has in position 191 either valine or leucine. They correspond to two alleles (character-determining portions) of a gene; the two types are called allotypes. The valine-containing genetic type has been designated as InV(a+), the leucine-containing type as InV(b+). Many more allotypes, called Gm allotypes, have been found in the gamma chains of the human IgG immunoglobulins; more than 20 Gm allotypes are known. Certain combinations of Gm types occur. For example, the combination of Gm types 5, 6, and 11 has been found in Caucasians and African Americans but not in Chinese; the combination of 1, 2, and 17 has not been found in African Americans; and the combination of 1, 4, and 17 has not been found in Caucasians. Allotypes have also been discovered to occur in a number of other animals, including rabbits and mice.

It is understandable from the occurrence of a large number of allotypes that antibodies, even if produced in response to a single antigen, are mixtures of different allotypes. The existence of several classes of antibodies, of different allotypes, and of adaptation of the variable portions of antibodies to different regions of an antigen molecule results in a multiplicity of antibody molecules even if only a single antigen is administered. For this reason it has not yet been possible to unravel the amino acid sequence in the variable portion of antibody molecules. Much of the amino acid sequence in the constant regions of the *L* and *H* chains of humans and rabbit immunoglobulins, however, has been resolved.

Enzymes

Practically all of the numerous and complex biochemical reactions that take place in animals, plants, and microorganisms are regulated by enzymes. These catalytic proteins are efficient and specific—that is, they accelerate the rate of one kind of chemical reaction of one type of compound, and they do so in a far more efficient manner than human-made catalysts. They are controlled by activators and inhibitors that initiate or block reactions. All cells contain enzymes, which usually vary in number and composition, depending on the cell type; an average mammalian cell, for example, is approximately one one-billionth (10^{-9}) the size of a drop of water and generally contains about 3,000 enzymes.

The existence of enzymes was established in the middle of the 19th century by scientists studying the process of fermentation. The discovery of the role of enzymes as catalysts followed rapidly. Developments before 1850 included (in 1833) the separation from malt of the enzyme amylase, which converts starch into sugar, and (in 1836) the isolation from the stomach wall of animals of a component of gastric juice that could partially digest food in a test tube, the enzyme pepsin.

Enzymes were known for many years as *ferments*, a term derived from the Latin word for yeast. In 1878 the name *enzyme*, from the Greek words meaning "in yeast," was introduced; since the late 19th century it has been employed universally.

Role of Enzymes in Metabolism

Some enzymes help to break down large nutrient molecules, such as proteins, fats, and carbohydrates, into smaller molecules. This process occurs during the digestion of foodstuffs in the stomach and intestines of animals. Other enzymes guide the smaller, broken-down molecules through the intestinal wall into the bloodstream. Still other enzymes promote the formation of large, complex molecules from the small, simple ones to produce cellular constituents. Enzymes are also responsible for numerous other functions, which include the storage and release of energy, the course of reproduction, the processes of respiration, and vision. They are indispensable to life.

Each enzyme is able to promote only one type of chemical reaction. The compounds on which the enzyme acts are called substrates. Enzymes operate in tightly organized metabolic systems called pathways. A seemingly simple biological phenomenon—the contraction of a muscle, for example, or the transmission of a nerve impulse—actually involves a large number of chemical steps in which one or more chemical compounds (substrates) are converted to substances called products; the product of one step in a metabolic pathway serves as the substrate for the succeeding step in the pathway.

The role of enzymes in metabolic pathways can be illustrated diagrammatically. The chemical compound represented by A is converted to product E in a series of enzyme-catalyzed steps, in which intermediate compounds represented by B, C, and D are formed in succession. They act as substrates for enzymes represented by 2, 3, and 4. Compound A may also be converted by another series of steps, some of which are the same as those in the pathway for the formation of E, to products represented by G and H.

$$
\begin{array}{c}
A \\
1\ \big| \\
B \\
2\ \big| \\
C \\
3\ \diagup\ \ 5\diagdown \quad \overset{H}{\diagup}7 \\
D \qquad F \\
4\ \big|\qquad \big|6 \\
E \qquad G
\end{array}
$$

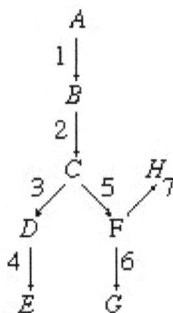

The letters represent chemical compounds; numbers represent enzymes that catalyze individual reactions. The relative heights represent the thermodynamic energy of the compounds (e.g., compound A is more energy-rich than B, B more energy-rich than C). Compounds A, B, etc., change very slowly in the absence of a catalyst but do so rapidly in the presence of catalysts 1, 2, 3, etc.

The regulatory role of enzymes in metabolic pathways can be clarified by using a simple analogy: that between the compounds, represented by letters in the diagram, and a series of connected water reservoirs on a slope. Similarly, the enzymes represented by the numbers are analogous to the valves of the reservoir system. The valves control the flow of water in the reservoir; that is, if only valves 1, 2, 3, and 4 are open, the water in A flows only to E, but, if valves 1, 2, 5, and 6 are open, the water in A flows to G. In a similar manner, if enzymes 1, 2, 3, and 4 in the metabolic pathway are active, product E is formed, and, if enzymes 1, 2, 5, and 6 are active, product G is formed. The activity or lack of activity of the enzymes in the pathway therefore determines the fate of compound A; i.e., it either remains unchanged or is converted to one or more products. In addition, if products are formed, the activity of enzymes 3 and 4 relative to that of enzymes 5 and 6 determines the quantity of product E formed compared with product G.

Both the flow of water and the activity of enzymes obey the laws of thermodynamics; hence, water in reservoir F cannot flow freely to H by opening valve 7, because water cannot flow uphill. If, however, valves 1, 2, 5, and 7 are open, water flows from F to H, because the energy conserved during the downhill flow of water through valves 1, 2, and 5 is sufficient to allow it to force the water up through valve 7. In a similar way, enzymes in the metabolic pathway cannot convert compound F directly to H unless energy is available; enzymes are able to utilize energy from energy-conserving reactions in order to catalyze reactions that require energy. During the enzyme-catalyzed oxidation of carbohydrates to carbon dioxide and water, energy is conserved in the form of an energy-rich compound, adenosine triphosphate (ATP). The energy in ATP is utilized during an energy-consuming process such as the enzyme-catalyzed contraction of muscle.

Because the needs of cells and organisms vary, not only the activity but also the synthesis of enzymes must be regulated; e.g., the enzymes responsible for muscular activity in a leg muscle must be activated and inhibited at appropriate times. Some cells do not need certain enzymes; a liver cell, for example, does not need a muscle enzyme. A bacterium does not need enzymes to metabolize substances that are not present in its growth medium. Some enzymes, therefore, are not formed in certain cells, others are synthesized only when required, and still others are found in all cells. The formation and activity of enzymes are regulated not only by genetic mechanisms but also by organic secretions (hormones) from endocrine glands and by nerve impulses. Small molecules also play an important role.

If an enzyme is defective in some respect, disease may occur. The enzymes represented by the numbers 1 to 4 in the diagram must function during the conversion of the starting substance A to the product E. If one step is blocked because an enzyme is unable to function, product E may not be formed; if E is necessary for some vital function, disease results. Many inherited diseases and conditions of humans result from a deficiency of one enzyme. Some of these are listed in the table. Albinism, for example, results from an inherited lack of ability to synthesize the enzyme tyrosinase, which catalyzes one step in the pathway by which the pigment for hair and eye colour is formed.

Enzymes identified with hereditary diseases	
Disease name	Defective enzyme
Albinism	Tyrosinase
Phenylketonuria	Phenylalanine hydroxylase
Fructosuria	Fructokinase
Methemoglobinemia	Methemoglobin reductase
Galactosemia	Galactose-1-phosphate uridyl transferase

Other Functions

Enzymes play an increasingly important role in medicine. The enzyme thrombin is used to promote the healing of wounds. Other enzymes are used to diagnose certain kinds of disease, to cause the remission of some forms of leukemia—a disease of the blood-forming organs—and to counteract unfavourable reactions in people who are allergic to penicillin. The enzyme lysozyme, which destroys cell walls, is used to kill bacteria. Enzymes have also been investigated for their potential to prevent tooth decay and to serve as anticoagulants in the treatment of thrombosis, a disease characterized by the formation of a clot, or plug, in a blood vessel. Enzymes may eventually be used to control enzyme deficiencies and abnormalities resulting from diseases.

It might also be noted in passing that enzymes are used in industrial processes involving the preparation of certain chemical compounds and the tanning of leather. They also are valuable in analytical procedures involving the detection of very small quantities of specific substances. Enzymes are necessary in various food-related industries, including cheese making, the brewing of beer, the aging of wine, and the baking of bread. Enzymes also may be used to clean clothes.

General Properties

Classification and Nomenclature

The first enzyme name, proposed in 1833, was diastase. Sixty-five years later, French microbiologist and chemist Émile Duclaux suggested that all enzymes be named by adding -ase to a root indicative of the nature of the substrate of the enzyme. Although enzymes are no longer named in such a simple manner, with the exception of a few—e.g., pepsin, trypsin, chymotrypsin, papain—most enzyme names do end in -ase.

Any systematic classification of enzymes should be based on a common property or quality that varies sufficiently to be useful as a distinguishing feature. In this regard, three properties of enzymes could serve as a basis for enzyme classification—the exact chemical nature of the enzyme, the chemical nature of the substrate, and the nature of the reaction catalyzed. In addition, although, as indicated above, early attempts at enzyme classification were based on the nature of broad groups of substrates (e.g., enzymes called carbohydrases act on carbohydrates), close functional similarities among enzymes in different groups were often obscured. By general agreement, enzymes now are classified according to their substrates and the nature of the reaction they catalyze.

In an attempt to devise a rational system of enzyme nomenclature, two names are given to an enzyme. One, known as the systematic name, is based on logical principles but is often long and awkward; the other, "trivial" name is short and generally used but not usually exact or systematic. In the scheme of systematic nomenclature, six main groups of enzymatic reactions are recognized; each catalyzes one reaction type and is subdivided on the basis of detailed definitions of the reaction catalyzed and of the substrate involved in the reaction. Enzymes that catalyze reactions in which hydrogen is transferred belong to the group known as oxidoreductases; those that catalyze the introduction of the elements of water at a specific site in a molecule are called hydrolases. The other four groups of reactions are the transferases—which catalyze reactions in which substances other than hydrogen are transferred—the lyases, the isomerases, and the ligases. Oxidoreductases and transferases account for about 50 percent of the approximately 1,000 enzymes recognized thus far. The table lists a few enzymes, their trivial names, their systematic names, and their biological roles.

Classification of Some Enzymes

Systematic name*		Trivial name	Reaction catalyzed	Biological role
Code number**	Name***			
1.1.1.1	Alcohol: NAD oxidoreductase	Alcohol dehydrogenase	Alcohol + NAD → acetaldehyde NADH	Alcoholic fermentation
1.1.1.27	L-lactate: NAD oxidoreductase	Lactic dehydrogenase	Lactate + NAD → pyruvate + NADH	Carbohydrate metabolism
2.7.1.40	ATP: pyruvate phosphotransferase	Pyruvate kinase	Pyruvic acid + ATP → phosphoenolpyruvic acid + ADP	Carbohydrate metabolism
3.1.1.7	Acetylcholine: acetylhydrolase	Acetylcholinesterase	Acetylcholine + H_2O → acetate + choline	Nerve-impulse conduction

*Based on recommendations (1964) of the International Union of Biochemistry.

**The numbering system is as follows: the first number places the enzyme in one of six general groups—1, oxidoreductases; 2, transferases; 3, hydrolases; 4, lyases; 5, iomerases; and 6, ligases. The second number places the enzyme in a subclass based on substrate type or reaction type; e.g., the enzyme may act on molecules with –CHOH groups. The third number places the enzyme in a subsubclass, which specifies the reaction type more fully; e.g., NAD coenzyme required. The fourth number is the serial number of the enzyme in its subsubclass.

***NAD and NADH represent the oxidized and reduced forms of nicotinamide adenine dinucleotide (NAD), respectively; ATP and ADP represent adenosine triphosphate and adenosine diphosphate, respectively.

Chemical Nature

Little was known about the chemical nature of enzymes until the beginning of the 20th century, although scientists were almost convinced that they were proteins. In 1926 the enzyme urease was the first to be crystallized and clearly identified as a protein. Within the next few years the digestive enzymes pepsin, trypsin, and chymotrypsin were shown to be proteins. Since that time hundreds of enzymes, all of them proteins, have been prepared and characterized by chemical methods. Much of the knowledge of protein chemistry has, in fact, resulted from studies involving enzymes and from attempts to understand their nature and mode of action.

Although some enzymes consist of a single chain of the amino acids (i.e., simple organic molecules containing nitrogen), most enzymes are composed of more than one chain. Each chain is called a subunit. Many enzymes have two, four, or six subunits, and some consist of as many as 12 to 60 subunits. In many cases the subunits have identical structures; in others, however, several different types of subunit chains are involved.

With the exception of proteins that act as structural elements, most of the proteins in physiologically active tissues such as kidney and liver are enzymes. Regardless of the exact amount of enzymatic protein in an organism, it is clear that hundreds of different enzymes must be present in each tissue to account for the myriad reactions composing metabolism.

Cofactors

Although some enzymes consist only of protein, many are complex proteins; i.e., they have a protein component and a so-called cofactor. A complete enzyme is called a holoenzyme; if the cofactor is removed, the protein, no longer enzymatically active, is called the apoenzyme. A cofactor may be a metal—such as iron, copper, or magnesium—a moderately sized organic molecule called a prosthetic group, or a special type of substrate molecule known as a coenzyme. The cofactor may aid in the catalytic function of an enzyme, as do metals and prosthetic groups, or take part in the enzymatic reaction, as do coenzymes.

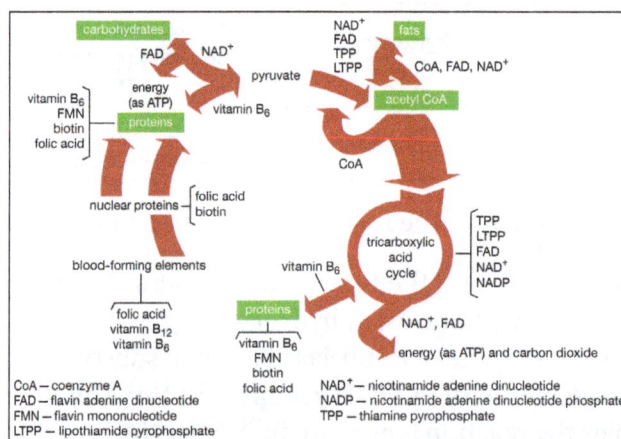

Functions of B-vitamin coenzymes in metabolism.

A coenzyme serves as a type of substrate in certain enzymatic reactions and thus reacts in the exact proportions (i.e., stoichiometrically) required for reaction, rather than in catalytic quantities. A coenzyme may, for example, assume the role of a hydrogen acceptor, as does nicotinamide adenine dinucleotide (NAD), which accepts hydrogen from the substrate, or a chemical-group donor, as does adenosine triphosphate (ATP), which donates phosphoric acid to the substrate. After ATP has donated a phosphoric acid molecule to the substrate, the phosphoric acid can be reacquired in a second stoichiometric reaction catalyzed by a second enzyme. The catalytic nature of a coenzyme is apparent only when it couples the activities of two enzymes in this way. Coenzymes thus are the links, or shuttles, in metabolic pathways that enable substances—e.g., hydrogen, phosphoric acid—to be exchanged.

The Nature of Enzyme-catalyzed Reactions

The Nature of Catalysis

In a chemical reaction—for example, one in which substance A is converted into product B—a point of equilibrium eventually is reached at which no further chemical change occurs; i.e., the rate of conversion of A to B equals the rate of conversion of B to A. The so-called thermodynamic-equilibrium constant expresses this chemical equilibrium. A catalyst may be defined as a substance that accelerates a chemical reaction but is not consumed in the process. The amount of catalyst has no relationship to the quantity of substance altered; very small amounts of enzymes are very efficient catalysts. Because the presence of an enzyme accelerates the rate of conversion of a compound to a product, it accelerates the approach to equilibrium; it does not, however, influence the equilibrium point attained.

The molecules in the watery medium of the cell are in constant thermal motion but, because they are more or less stable compounds, they would react only occasionally to form products in the absence of enzymes. There exists an energy barrier to the reaction of a molecule. The energy required to overcome the barrier to reaction is called the energy of activation. A reaction proceeds to equilibrium only if the molecules have sufficient energy of activation to form an activated complex, from which products can be derived. Enzymes greatly increase the chances for reactions by their ability to make large numbers of specific molecules more reactive (i.e., unstable) by forming intermediate compounds with them. The unstable intermediates quickly break down to form stable products, and the enzymes, unchanged by the reaction, are able to catalyze the formation of additional products.

The Role of the Active Site

That the compound on which an enzyme acts (substrate) must combine in some way with it before catalysis can proceed is an old idea, now supported by much experimental evidence. The combination of substrate molecules with enzymes involves collisions between the two. Enzymes are large molecules, the molecular weights of which (based on the weight of a hydrogen atom as 1) range from several thousand to several million. The substrates on which enzymes act usually have molecular weights of several hundred. Because of the difference in size between the two, only a fraction of the enzyme is in contact with the substrate; the region of contact is called the active site. Usually, each subunit of an enzyme has one active site capable of binding substrate.

The characteristics of an enzyme derive from the sequence of amino acids, which determine the shape of the enzyme (i.e., the structure of the active site) and hence the specificity of the enzyme. The forces that attract the substrate to the surface of an enzyme may be of a physical or a chemical nature. Electrostatic bonds may occur between oppositely charged groups—the circles containing plus and minus signs on the enzyme are attracted to their opposites in the substrate molecule. Such electrostatic bonds can occur with groups that are completely positively or negatively charged (i.e., ionic groups) or with groups that are partially charged (i.e., dipoles). The attractive forces between substrate and enzyme may also involve so-called hydrophobic bonds, in which the oily, or hydrocarbon, portions of the enzyme (represented by H-labelled circles) and the substrate are forced together in the same way as oil droplets tend to coalesce in water.

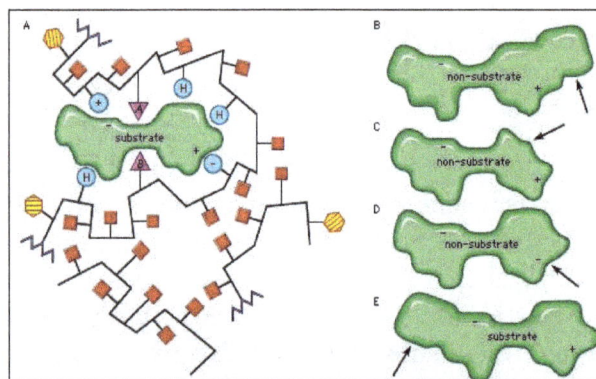

Enzyme; active site The role of the active site in the
lock-and-key fit of a substrate (the key) to an enzyme (the lock).

Modifications in the structure of the amino acids at or near the active site usually affect the enzyme's activity, because these amino acids are intimately involved in the fit and attraction of the substrate to the enzyme surface. The characteristics of the amino acids near the active site determine whether or not a substrate molecule will fit into the site. A molecule that is too bulky in the wrong places cannot fit into the active site and thus cannot react with the enzyme. In a similar manner, a molecule lacking essential attractive forces or the appropriately charged regions might not be bound to the enzyme. On the other hand, a molecule with a bulky group at a position such that it does not interfere with the binding of the molecule to the enzyme or with the function of the active site is able to serve as a substrate for the enzyme. The idea of a fit between substrate and enzyme, called the "key–lock" hypothesis, was proposed by German chemist Emil Fischer in 1899 and explains one of the most important features of enzymes, their specificity. In most of the enzymes studied thus far, a cleft, or indentation, into which the substrate fits is found at the active site.

The Specificity of Enzymes

Since the substrate must fit into the active site of the enzyme before catalysis can occur, only properly designed molecules can serve as substrates for a specific enzyme; in many cases, an enzyme will react with only one naturally occurring molecule. Two oxidoreductase enzymes will serve to illustrate the principle of enzyme specificity. One (alcohol dehydrogenase) acts on alcohol, the other (lactic dehydrogenase) on lactic acid; the activities of the two, even though both are oxidoreductase enzymes, are not interchangeable—i.e., alcohol dehydrogenase will not catalyze a reaction involving lactic acid or vice versa, because the structure of each substrate differs sufficiently to prevent its fitting into the active site of the alternative enzyme. Enzyme specificity is essential because it keeps separate the many pathways, involving hundreds of enzymes, that function during metabolism.

Not all enzymes are highly specific. Digestive enzymes such as pepsin and chymotrypsin, for example, are able to act on almost any protein, as they must if they are to act upon the varied types of proteins consumed as food. On the other hand, thrombin, which reacts only with the protein fibrinogen, is part of a very delicate blood-clotting mechanism and thus must act only on one compound in order to maintain the proper functioning of the system.

When enzymes were first studied, it was thought that most of them were "absolutely specific"—that they would react with only one compound. In most cases, however, a molecule other than the natural substrate can be synthesized in the laboratory; it is enough like the natural substrate to react with the enzyme. Use of these synthetic substrates has been valuable in understanding enzymatic action. It must be remembered, however, that, in the living cell, many enzymes are absolutely specific for the compounds found there.

All enzymes isolated thus far are specific for the type of chemical reaction they catalyze—i.e., oxidoreductases do not catalyze hydrolase reactions, and hydrolases do not catalyze reactions involving oxidation and reduction. An enzyme therefore catalyzes a specific chemical reaction but may be able to do so on several similar compounds.

The Mechanism of Enzymatic Action

An enzyme attracts substrates to its active site, catalyzes the chemical reaction by which products are formed, and then allows the products to dissociate (separate from the enzyme surface). The combination formed by an enzyme and its substrates is called the enzyme–substrate complex. When two substrates and one enzyme are involved, the complex is called a ternary complex; one substrate and one enzyme are called a binary complex. The substrates are attracted to the active site by electrostatic and hydrophobic forces, which are called noncovalent bonds because they are physical attractions and not chemical bonds.

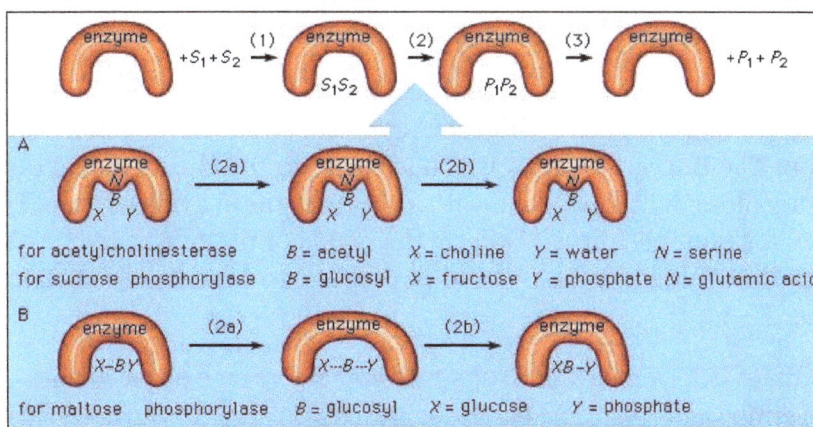

Mechanisms of enzymatic action.

As an example, assume two substrates (S_1 and S_2) bind to the active site of the enzyme during step 1 and react to form products (P_1 and P_2) during step 2. The products dissociate from the enzyme surface in step 3, releasing the enzyme. The enzyme, unchanged by the reaction, is able to react with additional substrate molecules in this manner many times per second to form products. The step in which the actual chemical transformation occurs is of great interest, and, although much is known about it, it is not yet fully understood. In general there are two types of enzymatic mechanisms, one in which a so-called covalent intermediate forms and one in which none forms.

In the mechanism by which a covalent intermediate—i.e., an intermediate with a chemical bond between substrate and enzyme—forms, one substrate, B−X, for example, reacts with the group N on the enzyme surface to form an enzyme-B intermediate compound. The intermediate compound then reacts with the second substrate, Y, to form the products B−Y and X.

Many enzymes catalyze reactions by this type of mechanism. Acetylcholinesterase is used as a specific example in the sequence described below. The two substrates (S_1 and S_2) for acetylcholinesterase are acetylcholine (i.e., B−X) and water (Y). After acetylcholine (B−X) binds to the enzyme surface, a chemical bond forms between the acetyl moiety (B) of acetylcholine and the group N (part of the amino acid serine) on the enzyme surface. The result of the formation of this bond, called an acyl−serine bond, is one product, choline (X), and the enzyme-B intermediate compound (an acetyl−enzyme complex). The water molecule (Y) then reacts with the acyl−serine bond to form the second product, acetic acid (B−Y), which dissociates from the enzyme. Acetylcholinesterase is regenerated and is again able to react with another molecule of acetylcholine. This kind of reaction, involving the formation of an intermediate compound on the enzyme surface, is generally called a double displacement reaction.

Sucrose phosphorylase acts in a similar way. The substrate for sucrose phosphorylase is sucrose, or glucosyl-fructose (B−X), and the group N on the enzyme surface is a chemical group called a carboxyl group (COOH). The enzyme-B intermediate, a glucosyl−carboxyl compound, reacts with phosphate (Y) to form glucosyl-phosphate (B−Y). The other product (X) is fructose.

In double displacement reactions, the covalent intermediate between enzyme and substrate apparently influences the reaction to proceed more rapidly. Because the enzyme is unaltered at the end of the reaction, it functions as a true catalyst, even though it is temporarily altered during the enzymatic process.

Although many enzymes form a covalent intermediate, the mechanism is not essential for catalysis. One substrate (Y) reacts directly with the second substrate (X−B), in a so-called single displacement reaction. The B moiety, which is transformed in the chemical reaction, is involved in only one reaction and does not form a bond with a group on the enzyme surface. The enzyme maltose phosphorylase, for example, directly affects the bonds of the substrates (B−X and X), which, in this case, are maltose (glucosylglucose) and phosphate, to form the products, glucose (X) and glucosylphosphate (B−Y).

Covalent intermediates between part of a substrate and an enzyme occur in many enzymatic reactions, and various amino acids—serine, cysteine, lysine, and glutamic acid—are involved.

The Rate of Enzymatic Reactions

The Michaelis-Menten Hypothesis

If the velocity of an enzymatic reaction is represented graphically as a function of the substrate concentration (S), the curve obtained in most cases is a hyperbola. The mathematical expression of this curve, shown in the equation below, was developed in 1912–13 by German biochemists Leonor

Michaelis and Maud Leonora Menten. In the equation, VM is the maximal velocity of the reaction, and KM is called the Michaelis constant.

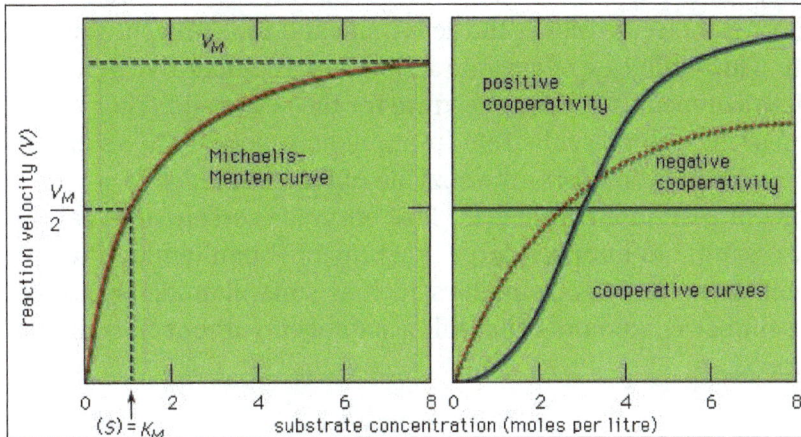

Curves representing enzyme action.

$$velocity = \frac{V_M(S)}{K_M + (S)}.$$

The shape of the curve is a logical consequence of the active-site concept; i.e., the curve flattens at the maximum velocity (VM), which occurs when all the active sites of the enzyme are filled with substrate. The fact that the velocity approaches a maximum at high substrate concentrations provides support for the assumption that an intermediate enzyme–substrate complex forms. At the point of half the maximum velocity, the substrate concentration in moles per litre (M) is equal to the Michaelis constant, which is a rough measure of the affinity of the substrate molecule for the surface of the enzyme. KM values usually vary from about 10^{-8} to 10^{-2} M, and VM from 10^5 to 10^9 molecules of product formed per molecule of enzyme per second. The value for VM is referred to as the turnover number when expressed as moles of product formed per mole of enzyme per minute. The binding of molecules that inhibit or activate the protein surface usually results in similar types.

Enzymes are more efficient than human-made catalysts operating under the same conditions. Because many enzymes with different specificities occur in a cell, adequate space exists only for a few enzyme molecules catalyzing one specific reaction. Each enzyme, therefore, must be very efficient. One molecule of the enzyme catalase, for example, can produce 10^{12} molecules of oxygen per second. The catalytic groups at the active site of an enzyme act 10^6 to 10^9 times more effectively than do analogous groups in a nonenzymatic reaction.

The reason for the great efficiency of enzymes is not completely understood. It results in part from the precise positioning of the substrates and the catalytic groups at the active site, which serves to increase the probability of collision between the reacting atoms. In addition, the environment at the active site may be favourable for reaction—that is, acidic and basic groups may act together more effectively there, or some strain may be induced in the substrate molecules so that their bonds are broken more easily, or the orientation of the reacting substrates may be optimal at the enzyme surface. The theories that have been formulated to account for the high catalytic efficiency of enzymes, although reasonable, still remain to be proved.

Inhibition of Enzymes

Some molecules very similar to the substrate for an enzyme may be bound to the active site but be unable to react. Such molecules cover the active site and thus prevent the binding of the actual substrate to the site. This inhibition of enzyme action is of a competitive nature, because the inhibitor molecule actually competes with the substrate for the active site. The inhibitor sulfanilamide, for example, is similar enough to a substrate (p-aminobenzoic acid) of an enzyme involved in the metabolism of folic acid that it binds to the enzyme but cannot react. It covers the active site and prevents the binding of p-aminobenzoic acid. This enzyme is essential in certain disease-causing bacteria but is not essential to humans; large amounts of sulfanilamide therefore kill the microorganism but do not harm humans. Inhibitors such as sulfanilamide are called antimetabolites. Sulfanilamide and similar compounds that kill a pathogen without harming its host are widely used in chemotherapy.

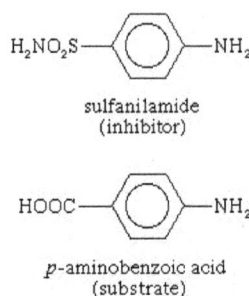

$$H_2NO_2S-\bigcirc-NH_2$$

sulfanilamide
(inhibitor)

$$HOOC-\bigcirc-NH_2$$

p-aminobenzoic acid
(substrate)

Some inhibitors prevent, or block, enzymatic action by reacting with groups at the active site. The nerve gas diisopropyl fluorophosphate, for example, reacts with the serine at the active site of acetylcholinesterase to form a covalent bond. The nerve gas molecule involved in bond formation prevents the active site from binding the substrate, acetylcholine, thereby blocking catalysis and nerve action. Iodoacetic acid similarly blocks a key enzyme in muscle action by forming a bulky group on the amino acid cysteine, which is found at the enzyme's active site. This process is called irreversible inhibition.

Some inhibitors modify amino acids other than those at the active site, resulting in loss of enzymatic activity. The inhibitor causes changes in the shape of the active site. Some amino acids other than those at the active site, however, can be modified without affecting the structure of the active site; in these cases, enzymatic action is not affected.

Such chemical changes parallel natural mutations. Inherited diseases frequently result from a change in an amino acid at the active site of an enzyme, thus making the enzyme defective. In some cases, an amino acid change alters the shape of the active site to the extent that it can no longer react; such diseases are usually fatal. In others, however, a partially defective enzyme is formed, and an individual may be very sick but able to live.

Effects of Temperature

Enzymes function most efficiently within a physiological temperature range. Since enzymes are protein molecules, they can be destroyed by high temperatures. An example of such destruction, called protein denaturation, is the curdling of milk when it is boiled. Increasing temperature has two effects on an enzyme: First, the velocity of the reaction increases somewhat, because the rate of chemical reactions tends to increase with temperature; and, second, the enzyme is increasingly denatured. Increasing temperature thus increases the metabolic rate only within a limited range. If the temperature becomes too high, enzyme denaturation destroys life. Low temperatures also change the shapes of enzymes. With enzymes that are cold-sensitive, the change causes loss of activity. Both excessive cold and heat are therefore damaging to enzymes.

The degree of acidity or basicity of a solution, which is expressed as pH, also affects enzymes. As the acidity of a solution changes—i.e., the pH is altered—a point of optimum acidity occurs, at which the enzyme acts most efficiently. Although this pH optimum varies with temperature and is influenced by other constituents of the solution containing the enzyme, it is a characteristic property of enzymes. Because enzymes are sensitive to changes in acidity, most living systems are highly buffered; i.e., they have mechanisms that enable them to maintain a constant acidity. This acidity level, or pH, is about 7 in most organisms. Some bacteria function under moderately acidic or basic conditions; and the digestive enzyme pepsin acts in the acid milieu of the stomach.

Enzyme Flexibility and Allosteric Control

The Induced-fit Theory

The key–lock hypothesis does not fully account for enzymatic action; i.e., certain properties of enzymes cannot be accounted for by the simple relationship between enzyme and substrate proposed by the key–lock hypothesis. A theory called the induced-fit theory retains the key–lock idea of a fit of the substrate at the active site but postulates in addition that the substrate must do more than simply fit into the already preformed shape of an active site. Rather, the theory states, the binding of the substrate to the enzyme must cause a change in the shape of the enzyme that results in the proper alignment of the catalytic groups on its surface. This concept has been likened to the fit of a hand in a glove, the hand (substrate) inducing a change in the shape of the glove (enzyme). Although some enzymes appear to function according to the older key–lock hypothesis, most apparently function according to the induced-fit theory.

Typically, the substrate approaches the enzyme surface and induces a change in its shape that results in the correct alignment of the catalytic groups. In the case of the digestive enzyme carboxypeptidase, for example, the binding of the substrate causes a tyrosine molecule at the active site to move by as much as 15 angstroms. The catalytic groups at the active site react with the substrate to form products. The products separate from the enzyme surface, and the enzyme is able to repeat the sequence. Nonsubstrate molecules that are too bulky or too small alter the shape of the enzyme so that a misalignment of catalytic groups occurs; such molecules are not able to react even if they are attracted to the active site.

The induced-fit theory explains a number of anomalous properties of enzymes. An example is "noncompetitive inhibition," in which a compound inhibits the reaction of an enzyme but does not prevent the binding of the substrate. In this case, the inhibitor compound attracts the binding group so that the catalytic group is too far away from the substrate to react. The site at which the inhibitor binds to the enzyme is not the active site and is called an allosteric site. The inhibitor changes the shape of the active site to prevent catalysis without preventing binding of the substrate.

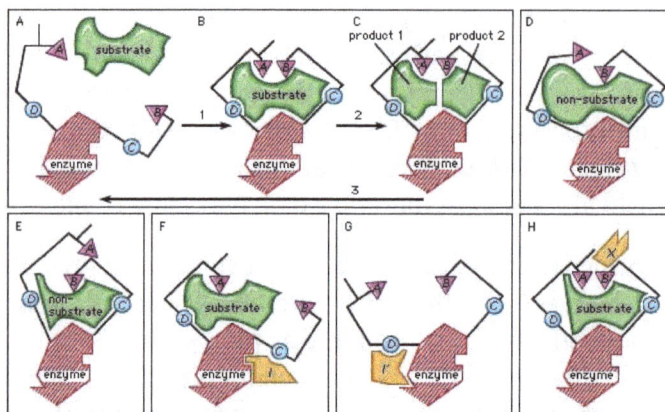

Induced-fit binding of a substrate to an enzyme surface and allosteric effects.

An inhibitor also can distort the active site by affecting the essential binding group; as a result, the enzyme can no longer attract the substrate. A so-called activator molecule affects the active site so that a nonsubstrate molecule is properly aligned and hence can react with the enzyme. Such activators can affect both binding and catalytic groups at the active site.

Enzyme flexibility is extremely important because it provides a mechanism for regulating enzymatic activity. The orientation at the active site can be disrupted by the binding of an inhibitor at a site other than the active site. Moreover, the enzyme can be activated by molecules that induce a proper alignment of the active site for a substrate that alone cannot induce this alignment.

The sites that bind inhibitors and activators are called allosteric sites to distinguish them from active sites. Allosteric sites are in fact regulatory sites able to activate or inhibit enzymatic activity by influencing the shape of the enzyme. When the activator or inhibitor dissociates from the enzyme, it returns to its normal shape. Thus, the flexibility of the protein structure allows the operation of a simple, reversible control system similar to a thermostat.

Types of Allosteric Control

Allosteric control can operate in many ways; two examples serve to illustrate some general effects. A pathway consisting of ten enzymes is involved in the synthesis of the amino acid histidine. When a cell contains enough histidine, synthesis stops—an appropriate economy move by the cell. Synthesis is stopped by the inhibition of the first enzyme in the pathway by the product, histidine. The inhibition of an enzyme by a product is called feedback inhibition; i.e., a product many steps removed from an initial enzyme blocks its action. Feedback inhibition occurs in many pathways in all living things.

Allosteric control can also be achieved by activators. The hormone adrenaline (epinephrine) acts in this way. When energy is needed, adrenaline is released and activates, by allosteric activation, the enzyme adenyl cyclase. This enzyme catalyzes a reaction in which the compound cyclic adenosine monophosphate (cyclic AMP) is formed from ATP. Cyclic AMP in turn acts as an allosteric activator of enzymes that speed the metabolism of carbohydrate to produce energy. This type of allosteric regulation also is widespread in biological systems. Thus, a combination of allosteric activation and inhibition allows the production of energy or materials when they are needed and shuts off production when the supply is adequate.

Allosteric control is a rapid method of regulating products continuously needed by living things. Yet some cells have no need for certain enzymes, and it would be wasteful for the cell to synthesize them. In this case, certain molecules, called repressors, prevent the synthesis of unneeded enzymes. The repressors are proteins that bind to DNA and prevent the first step in the process resulting in protein synthesis. If certain metabolites are added to cells that need an enzyme, enzyme synthesis occurs—i.e., it is induced. Addition of galactose to a growth medium containing Escherichia coli bacteria, for example, induces the synthesis of the enzyme beta-galactosidase. The bacteria thus can synthesize this galactose-metabolizing enzyme when it is needed and prevent its synthesis when it is not. The way in which the synthesis of enzymes is induced or repressed in mammalian systems is less understood but is believed to be similar.

Different types of cells in complex organisms have different enzymes, even though they have the same DNA content. The enzymes actually synthesized are the ones needed in a specific cell and vary not only for different types of cells—e.g., nerve, muscle, eye, and skin cells—but also for different species.

In an enzyme consisting of several subunits, or chains, alteration in the shape of one chain as a result of the influence either of a substrate molecule or of allosteric inhibitors or activators may change the shape of a neighbouring chain. As a result, the binding of a second molecule of substrate occurs in a different way from the binding of the first, and the third is different from the second. This phenomenon, called cooperativity, is characteristic of allosteric enzymes. Cooperativity is reflected by a sigmoid curve, as compared to the hyperbolic curve of Michaelis–Menten. An enzyme of several subunits that exhibits cooperativity is far more sensitive to control mechanisms than is an enzyme of one subunit and hence one active site.

The first example of cooperativity was observed in hemoglobin, which is not an enzyme but behaves like one in many ways. The absorption of oxygen in the lungs and its deposition in the tissues is far more efficient because the subunits of hemoglobin show positive cooperativity, so called because the first molecule of substrate makes it easier for the next to bind.

Negative cooperativity, in which the binding of one molecule makes it less easy for the next to bind, also occurs in living things. Negative cooperativity makes an enzyme less sensitive to fluctuations in concentrations of metabolites and may be important for enzymes that must be present in the cell at relatively constant levels of activity.

Some enzymes are closely associated aggregates of several enzyme units; the pyruvate dehydrogenase system, for example, contains five different enzymes, has a total molecular weight of 4,000,000, and consists of four different types of chains. Apparently, the enzymes in cells may be organized by forming complex units, by being absorbed on a cell wall, or by being isolated by membranes in special compartments. Since a pathway involves the stepwise modification of chemical compounds, aggregations of the enzymes in a given pathway facilitate their function in a manner similar to an industrial assembly line.

Protein Metabolism

Protein turnover is the balance between protein synthesis and protein degradation. Proteins are naturally occurring polymers made up of repeating units of 20 different amino acids and range from small peptide hormones of 8 to 10 residues to very large multi-chain complexes of several thousand amino acids. Protein synthesis occurs on ribosomes - large intracellular structures consisting of a small subunit (33 proteins, 1900 nucleotides of ribosomal RNA) and a large subunit (46 proteins, 4980 nucleotides of rRNA) - that move along the messenger RNA (mRNA) copy of the gene (DNA) that was transcribed. The process of protein synthesis is called translation where the mRNA is read in triplets (codons), each triplet directing the addition of an amino acid (via its specific transfer RNA (tRNA)) to the growing polypeptide chain.

The assembly of new proteins requires a source of amino acids which come from either the proteolytic breakdown (digestion) of proteins in the gastrointestinal tract or the degradation of proteins within the cell. Intracellular protein degradation is done by proteolytic enzymes called proteases and occurs generally in two cellular locations - lysosomes and proteosomes. Lysosomal proteases digest proteins of extracellular origin that have been taken up by the process of endocytosis. Proteosomes, which are large, barrel-shaped, ATP-dependent protein complexes, digest damaged or unneeded intracellular proteins that have been marked for destruction by the covalent attachment of chains of a small protein, ubiquitin.

In contrast to the situation with glucose and fatty acids, amino acids in excess of those needed for biosynthesis cannot be stored and are not excreted. Rather, surplus amino acids are used as metabolic fuel. Most of the amino groups of surplus amino acids are converted into urea through the urea cycle, whereas their carbon skeletons are transformed into acetyl CoA, acetoacetyl CoA, pyruvate, or one of the intermediates of the tricarboxylic acid cycle. Hence, fatty acids, ketone bodies and glucose can be formed from amino acids.

Protein Synthesis

The overall process of protein synthesis extends from gene transcription in the nucleus to polypeptide synthesis on ribosomes in the cytoplasm and is summarized in figure below.

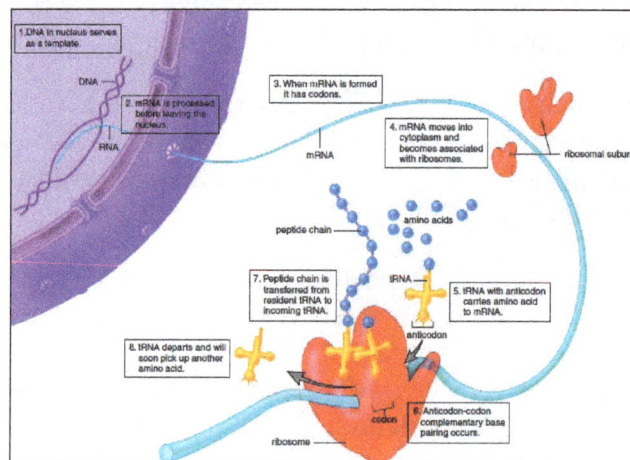

Genes and the Genetic Code

The human genome contains approximately 20,000 protein-coding genes that provide the instructions for the 250,000 to 1,000,000 proteins that operate in the human body over its lifetime. The number of proteins exceeds the number of genes because one gene can code for more than one protein and many proteins exist in multiple forms due to post-synthetic chemical modifications. In addition there are thousands of non-coding RNA genes that help regulate the protein coding genes.

Genetic information is stored in the nucleus of cells in the form of deoxyribose nucleic acid (DNA). DNA is composed of just four building blocks (bases) adenine (A), guanine (G), cytosine (C) and thymine (T) linked by a deoxyribose -phosphate backbone to form a double helix.

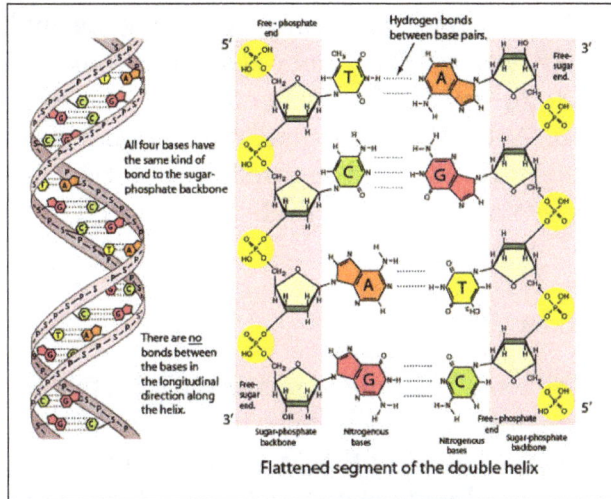

DNA structure showing the double helix and the base pairing (hydrogen bonds) where A pairs with T and G pairs with C.

As there are 20 amino acids, the code is not read as a single letter (4 possibilities only) or double letter (4x4 – 16 combinations) format, rather it is read in triplets called codons, with 4x4x4 (64) total combinations coding for all 20 amino acids as well as some punctuation (Stop/Start) instructions.

		Second Letter								
		U		C		A		G		
1st letter	U	UUU UUC	Phe	UCU UCC	Ser	UAU UAC	Tyr	UGU UGC	Cys	U C	3rd letter
		UUA UUG	Leu	UCA UCG		UAA UAG	Stop Stop	UGA UGG	Stop Trp	A G	
	C	CUU CUC	Leu	CCU CCC	Pro	CAU CAC	His	CGU CGC	Arg	U C	
		CUA CUG		CCA CCG		CAA CAG	Gln	CGA CGG		A G	
	A	AUU AUC	Ile	ACU ACC	Thr	AAU AAC	Asn	AGU AGC	Ser	U C	
		AUA AUG	Met	ACA ACG		AAA AAG	Lys	AGA AGG	Arg	A G	
	G	GUU GUC	Val	GCU GCC	Ala	GAU GAC	Asp	GGU GGC	Gly	U C	
		GUA GUG		GCA GCG		GAA GAG	Glu	GGA GGG		A G	

The genetic code.

The genetic sequence of DNA bases A, G, C, and T is copied and processed into the corresponding sequence of messenger RNA bases A, G, C and U for translation into polypeptide by the ribosome. The sequence of bases is read in triplets with sixty four possible combinations. Methionine (Met) and tryptophan (Trp) are each coded for by a single triplet (codon), asparagine (Asn), aspartic acid (Asp), glutamine (Gln), glutamic acid (Glu), cysteine (Cys), phenylalanine (Phe), tyrosine (Tyr) and lysine (Lys) each have two codons; isoleucine (Ile) has three; glycine (Gly), alanine (Ala), valine (Val), threonine (Thr) and proline (Pro) have four codons while serine (Ser) and arginine (Arg) have six. There are three stop codons that denote the C-terminus of the translated protein.

Transcription

Proteins are not synthesized directly from DNA, but from an RNA (ribose nucleic acid) copy derived from one strand of DNA by a process called transcription. Transcription occurs in the nucleus. The sections of the RNA gene copy that correspond to regions in the DNA that do not code for residues in the final protein (introns), are removed and the processed RNA copy - called messenger RNA - is transported to the cytoplasm where protein synthesis takes place. RNA contains three of the same bases as DNA, (adenine, guanine, cytosine) but employs uracil as the fourth base rather than thymine. Gene transcription is controlled by special DNA-binding proteins called transcription factors that are synthesized and activated/inactivated by protein hormones such as insulin and glucagon and non-protein hormones such as corticosteroids.

The Role of Transfer RNA

A transfer RNA (tRNA) is an adaptor molecule composed of RNA, typically 76 to 90 nucleotides in length. It serves as the physical link between the nucleotide sequence of the mRNA being translated and the resulting amino acid sequence of the synthesized protein. Thus tRNA is the means by which the correct amino acid - required to match each codon (triplet of bases) in the transcribed mRNA- is positioned in the peptidyl-transferase centre of the ribosome. One end of each tRNA contains a three-nucleotide sequence called the anticodon that can form three base pairs with a complementary three-nucleotide codon in mRNA during protein biosynthesis. Covalently attached to the other end of each tRNA is the amino acid that corresponds to the mRNA codon sequence. Each type of tRNA molecule can have only one type of amino acid attached to it and this is synthesized by enzymes called aminoacyl tRNA synthetases. One molecule of ATP is consumed in this process. Given the genetic code contains multiple codons that specify the same amino acid there will be several tRNA molecules bearing different anticodons which also carry the same amino acid.

Translation on Ribosomes

Protein synthesis is carried out in the cytoplasm by ribosomes - massive protein and RNA complexes that translate the nucleotide code on messenger RNA (mRNA) into functional protein. Eukaryotic organisms, which include humans, have two ribosomal subunits, the large 60S and small 40S, which combine to form the functional 80S complex. In contrast, prokaryotes such as bacteria have similar, but smaller subunits — a large 50S and small 30S, which combine to form a 70S complex.

Ribosomes have been the focus of structural and biochemical studies for more than 50 years and in 2000, Tom Steitz's laboratory at Yale University in Hartford Connecticut, published a high-resolution (2.4 Å) structure, of the large 50S subunit in the journal Science. At this resolution, the researchers were able to definitively place nearly all of the 50S subunit's 3,045 nucleotides and 31 proteins. The structure revealed that the ribosome is a ribozyme because the catalytic peptidyl transferase activity that catalyses peptide bond formation, linking the amino acids together in the growing peptide chain, is performed by the ribosomal RNA. Numerous initiation, elongation and release factors ensure that protein synthesis occurs progressively and with high specificity. In the past few years, high-resolution structures have provided molecular snapshots of different intermediates in ribosome-mediated translation in atomic detail. Together, these studies have revolutionized our understanding of the mechanism of protein synthesis. Tom Steitz shared the 2009 Nobel Prize for Chemistry with Venkatraman Ramakrishnan from the MRC Laboratory of Molecular

Biology, Cambridge, United Kingdom, and Ada E. Yonath from the Weizmann Institute of Science, Rehovot, Israel 'for their studies of the structure and function of the ribosome'.

The process of protein synthesis on ribosomes involves binding the mRNA in a tunnel formed between the two ribosomal subunits and initiating protein synthesis at the first codon. The two ribosomal subunits perform different roles in protein synthesis. The small ribosomal subunit mediates the correct inter¬actions between the anticodons of the tRNAs and the codons in the mRNA that they are translating in order to determine the order of the amino acids in the protein being synthesized. The large subunit contains the peptidyl-transferase centre (PTC), which catalyses the formation of peptide bonds in the growing polypeptide.

Both subunits contain three binding sites A, P and E, for tRNA molecules that are in three different functional states. The A site binds the aminoacyl-tRNA that is about to be incorporated into the growing polypeptide chain, the P site positions the peptidyl-tRNA (i.e. the tRNA with the growing peptide chain attached) and the E site is occupied by the deacylated tRNA before it dissociates from the ribosome (i.e. the tRNA after its attached peptide chain has been transferred (covalently linked) to the incoming amino acid on the aminoacyl tRNA).

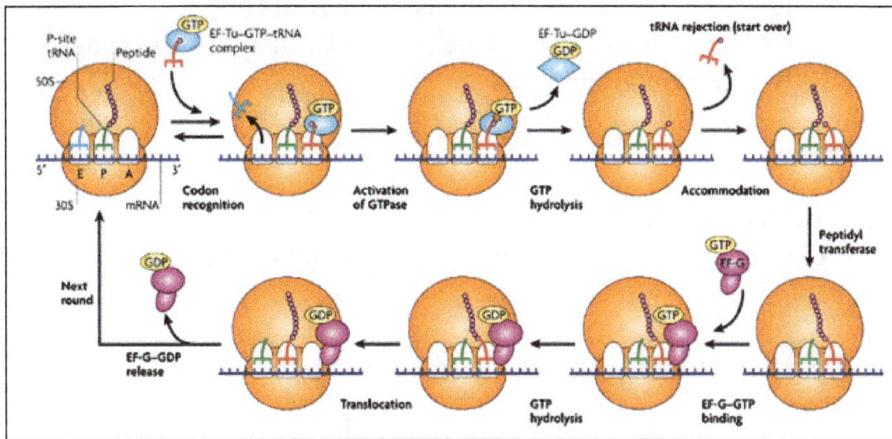

An overview of steps in protein synthesis. mRNA translation is initiated
with the binding of tRNAfmet to the P site.

  An incoming tRNA is delivered to the A site in complex with elongation factor (EF)-Tu–GTP. Correct codon–anticodon pairing activates the GTPase centre of the ribosome, which causes hydrolysis of GTP and release of the aminoacyl end of the tRNA from EF Tu. Binding of tRNA also induces conformational changes in ribosomal (r)RNA that optimally orientates the peptidyl-tRNA and aminoacyl-tRNA for the peptidyl-transferase reaction to occur, which involves the transfer of the peptide chain onto the A site tRNA. The ribosome must then shift in the 3' mRNA direction so that it can decode the next mRNA codon. Translocation of the tRNAs and mRNA is facilitated by binding of the GTPase EF G, which causes the deacylated tRNA at the P site to move to the E site and the peptidyl-tRNA at the A site to move to the P site upon GTP hydrolysis. The ribosome is then ready for the next round of elongation. The deacylated tRNA in the E site is released on binding of the next aminoacyl-tRNA to the A site. Elongation ends when a stop codon is reached, which initiates the termination reaction that releases the polypeptide.

Of central interest are the mechanisms of peptide bond formation and mRNA decoding, which are crucial processes in the elongation phase of protein synthesis by the ribosome. During this phase of protein synthesis, nascent polypeptides are elongated from the N- to the C-terminus by the addition of one amino acid at a time. This process is facilitated by two protein factors: Elongation factor Tu (EF-Tu), which facilitates the delivery of aminoacyl-tRNA to the A site of the ribosome, and elongation factor G (EF G), which promotes the translocation of the tRNAs and associated mRNA from their positions in the A and P sites to the P and E sites, respectively, and dissociates the previously bound E-site tRNA.

The accurate delivery of the correct aminoacyl-tRNA to the A site involves at least two distinct steps: (i) an interaction between the anticodon base triplet in the tRNA and the corresponding codon of the mRNA that resides in the A site of the ribosome and (ii) the communication of this correct formation of anticodon/codon Watson–Crick base pairing to the GTPase centre located in the large ribosomal subunit ~70 Å away, which results in the hydrolysis of the GTP bound to EF Tu. This GTP hydrolysis changes the conformation of EF-Tu resulting in its release from the tRNA and ribosome and the subsequent accommodation of the aminoacyl end of the tRNA into the peptidyl-transferase centre (PTC), which is followed rapidly by peptide bond formation.

At the end of the elongation cycle when the stop codon has been positioned in the A site, one of two protein release factors (RFs), RFI or RFII, binds to the A site and promotes the deacylation of the peptidyl-tRNA. A recycling factor, with the help of EF G, then leads to the dissociation of the release factor and the two ribosomal subunits.

Protein Catabolism

The assembly of new proteins requires a source of amino acids. These building blocks are generated by the digestion of proteins in the gastrointestinal tract and the degradation of proteins within the cell.

Protein Hydrolysis in the Digestive Tract

Protein digestion begins in the stomach, where the acidic environment favors protein denaturation. Denatured proteins are more accessible as substrates for proteolysis than are native proteins. The primary proteolytic enzyme of the stomach is the aspartate protease pepsin, a nonspecific protease that, remarkably, is maximally active at pH 2. Thus, pepsin can be active in the highly acidic environment of the stomach, even though other proteins undergo denaturation there. Protein degradation continues in the lumen of the intestine owing to the activity of proteolytic enzymes secreted by the pancreas. These are the serine proteases trypsin and chymotrypsin, and the carboxypeptidases (zinc metalloenzymes).

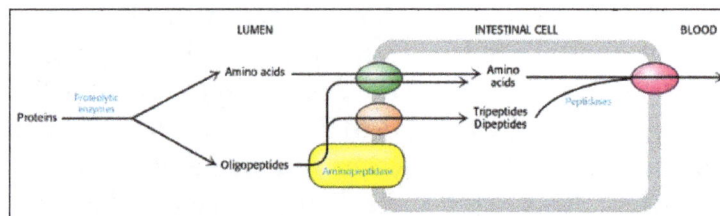

Digestion and absorption of proteins.

This battery of enzymes displays a wide array of specificity, and so the substrates are degraded into free amino acids as well as di- and tripeptides. Digestion is further enhanced by proteases, such as aminopeptidase N, that are located in the plasma membrane of the intestinal cells. Aminopeptidases digest proteins from the amino-terminal end. Single amino acids, as well as di- and tripeptides, are transported into the intestinal cells from the lumen and subsequently released into the blood for absorption by other tissues.

Protein Turnover and Intracellular Protein Breakdown

Protein turnover—the degradation and resynthesis of proteins—takes place constantly in cells. Protein degradation is as essential to the cell as protein synthesis. It is required to supply amino acids for fresh protein synthesis; to remove excess enzymes and to remove transcription factors that are no longer needed. Processes regulated by protein degradation include gene transcription, cell-cycle progression, organ formation, circadian rhythms, inflammatory responses, tumour suppression, cholesterol metabolism and antigen processing.

Although some proteins are very stable, many proteins are short lived, particularly those that are important in metabolic regulation. Altering the amounts of these proteins can rapidly change metabolic patterns. In addition, cells have mechanisms for detecting and removing damaged proteins. A significant proportion of newly synthesized protein molecules are defective because of errors in translation. Even proteins that are normal when first synthesized may undergo oxidative damage or be altered in other ways with the passage of time.

There are two major intracellular devices in which damaged or unneeded proteins are broken down: proteasomes and lysosomes.

Proteasomes deal primarily with endogenous proteins, i.e., proteins that were synthesized within the cell such as:

- Transcription factors.
- Cyclins (which must be destroyed to prepare for the next step in the cell cycle).
- Proteins encoded by viruses and other intracellular pathogens.
- Proteins that are folded incorrectly because of translation errors or because they are encoded by faulty genes (as in cystic fibrosis), or they have been damaged by other molecules in the cytosol.

Lysosomes deal primarily with:

- Extracellular proteins, e.g., plasma proteins, that are taken into the cell, e.g., by endocytosis.
- Cell-surface membrane proteins that are used in receptor-mediated endocytosis.
- Proteins (and other macromolecules) engulfed by autophagosomes.

Proteasomal Degradation of Poly-ubiquitinated Intracellular Proteins

Damaged or unneeded proteins are marked for destruction by the covalent attachment of chains of a small protein, ubiquitin. Polyubiquitinated proteins are subsequently degraded by a large, ATP-dependent complex called the proteasome. Proteasomes provide a controlled method for breaking down proteins safely within the environment of the cell. They chop obsolete or damaged proteins into small pieces, about 2 to 25 amino acids in length. Most of these are then completely broken down into amino acids by peptidases in the cell.

A simplified structure of the proteasome is shown in figure below.

The process of protein degradation in the proteasome involves the following stages.

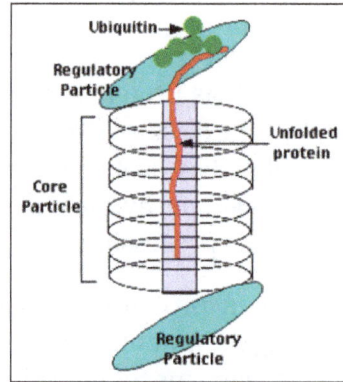

Cartoon of the proteasome.

It consists of a core particle made up of 2 copies of 14 different proteins, assembled into 2 rings of 7. The 4 rings are stacked on each other (like 4 doughnuts). There are two identical regulatory particles (RPs), one at each end of the core particle (CP). Each RP is made of 19 different proteins (none of them the same as those in the CP). Six of these are ATPases.

- Ubiquination: The proteasome is a multisubunit enzyme complex that plays a central role in the regulation of proteins that control cell-cycle progression and apoptosis, and has therefore become an important target for anticancer therapy. Before a protein is degraded, it is first flagged for destruction by the ubiquitin conjugation system, which ultimately results in the attachment of a polyubiquitin chain on the target protein. Ubiquination involves three enzymes, designated E1, E2 and E3. Initially, the terminal carboxyl group of ubiquitin is joined in a thioester bond to a cysteine residue on E1 (Ubiquitin-Activating Enzyme). This is an ATP-dependent step. The ubiquitin is then transferred to a sulfhydryl group on E2 (Ubiquitin-Conjugating Enzyme). E3, a Ubiquitin-Protein Ligase then promotes transfer of ubiquitin from E2 to the ε-amino group of a lysine residue of the targeted protein that has been recognized by that E3, forming an isopeptide bond. There are many distinct Ubiquitin ligases with differing substrate specificity. More ubiquitins may be added to form a chain of ubiquitins, the terminal carboxyl of each ubiquitin being linked to the ε-amino group of a lysine residue (Lys29 or Lys48) of the adjacent ubiquitin in the chain. A chain of four or more ubiquitin molecules targets proteins for degradation in proteasomes. Attachment of a single ubiquitin to a protein has other regulatory effects.

- Binding and denaturation: The proteasome's 19S regulatory cap at each end of the core, contains a large collection of regulatory subunits, (coloured blue in right-hand panel of figure), that recognize proteins that are tagged with ubiquitin and queued up for destruction. Once bound the protein is unfolded and denatured by a set of ATPases (colored magenta in right-hand panel of figure) using the energy of ATP. The unfolded protein is then translocated into the central cavity of the core particle.

- Hydrolysis: Since proteasomes perform their job inside a cell, figure below. surrounded by proteins, the protein-cutting ability of proteasomes is carefully controlled. The active sites are hidden away inside a cylindrical "core" particle, shown here in yellow and red in right-hand panel of Figure. The proteolytic core is composed of 2 inner β rings and 2 outer α rings. The 2 β rings each contain 3 proteolytic sites named for their trypsin-like, post-glutamyl peptide hydrolase-like (PGPH) (i.e., caspase-like), or chymotrypsin-like activity. These proteases on the inner surface of the two middle "doughnuts" break various specific peptide bonds of the denatured protein chain, producing a set of peptides averaging about 8 amino acids in length. These leave the core particle by an unknown route where they may be further broken down into individual amino acids by peptidases in the cytosol or in mammals may be incorporated in a class I histocompatibility molecule to be presented to the immune system as a potential antigen.

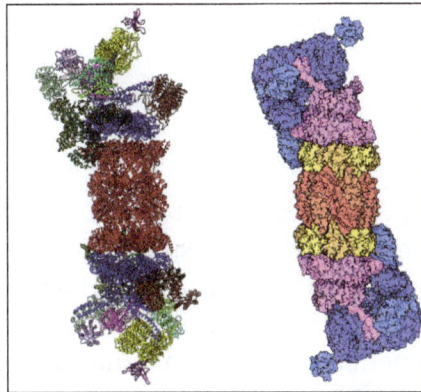

3D structure of the proteasome. Left panel is a ribbon diagram of the yeast 26S proteasome, right panel is a space filled model created by integrating several partial crystallographic structures into a near-atomic reconstruction of the proteasome obtained by analysis of 2.4 million images from electron microscopy.

- Ubiquitin release: As the denatured protein is hydrolysed the regulatory particle releases the pre-bound ubiquitins for reuse in the process.

Lysosomal Proteolysis (Endocytosed and Organellar Proteins)

Lysosomes contain a large variety of proteases that degrade proteins and other substances taken up by endocytosis. Lysosomal proteases include cathepsins (cysteine proteases), aspartate proteases and one zinc protease. Materials taken into a cell by inward budding of vesicles from the plasma membrane may be processed first in an endosomal compartment and then delivered into the lumen of a lysosome by fusion of a transport vesicle. Solute transporters embedded in the lysosomal membrane catalyze the exit of the products of lysosomal digestion (e.g., amino acids, sugars, cholesterol) to the cytosol. Lysosomes have a low internal pH due to activity of vacuolar ATPase, a H+ pump homologous to (but distinct from) the mitochondrial F1F0 ATPase. All intra-lysosomal hydrolases exhibit acidic pH optima.

Caspases and Apoptosis

Caspases are cysteine proteases involved in the activation and implementation of apoptosis (programmed cell death). Caspases get their name from the fact that they cleave on the carboxyl side of aspartic acid.

Caspases are a family of endoproteases that provide critical links in cell regulatory networks controlling inflammation and cell death. The activation of these enzymes is tightly controlled by their production as inactive zymogens that gain catalytic activity following signaling events promoting their aggregation into dimers or macromolecular complexes. Activation of apoptotic caspases results in inactivation or activation of substrates, and the generation of a cascade of signaling events permitting the controlled demolition of cellular components.

Activation of inflammatory caspases results in the production of active pro-inflammatory cytokines and the promotion of innate immune responses to various internal and external insults. Dysregulation of caspases underlies human diseases including cancer and inflammatory disorders, and major efforts to design better therapies for these diseases seek to understand how these enzymes work and how they can be controlled. Apoptosis and caspase function is one of the most heavily researched fields in molecular medicine.

Proteome

A proteome is the complete set of proteins expressed by an organism. The term can also be used to describe the assortment of proteins produced at a specific time in a particular cell or tissue type. The proteome is an expression of an organism's genome. However, in contrast with the genome, which is characterized by its stability, the proteome actively changes in response to various factors, including the organism's developmental stage and both internal and external conditions.

The study of the proteome is called proteomics, and it involves understanding how proteins function and interact with one another. For instance, many proteins fold into elaborate three-dimensional structures, and some form complexes with each other to perform their functions. In addition, proteins undergo modifications, which may occur either before or after translation. The proteome can be studied using a variety of techniques. For example, two-dimensional gel electrophoresis can be used to separate proteins by their sizes and by their charges. The proteome can also be studied using another laboratory technique called mass spectrometry, which identifies specific proteins within complex samples.

The term has been applied to several different types of biological systems. A cellular proteome is the collection of proteins found in a particular cell type under a particular set of environmental conditions such as exposure to hormone stimulation. It can also be useful to consider an organism's complete proteome, which can be conceptualized as the complete set of proteins from all of the various cellular proteomes. This is very roughly the protein equivalent of the genome. The term "proteome" has also been used to refer to the collection of proteins in certain sub-cellular biological systems. For example, all of the proteins in a virus can be called a viral proteome.

Marc Wilkins coined the term proteome in 1994 in a symposium on "2D Electrophoresis: from protein maps to genomes" held in Siena in Italy. It appeared in print in 1995, with the publication of part of his PhD thesis. Wilkins used the term to describe the entire complement of proteins expressed by a genome, cell, tissue or organism.

Size and Contents

The proteome can be larger than the genome, especially in eukaryotes, as more than one protein can be produced from one gene due to alternative splicing (e.g. human proteome consists 92,179 proteins out of which 71,173 are splicing variants). On the other hand, not all genes are translated to proteins, and many known genes encode only RNA which is the final functional product. Moreover, complete proteome size vary depending the kingdom of life. For instance, eukaryotes, bacteria, archaea and viruses have on average 15,145, 3,200, 2,358 and 42 proteins respectively encoded in their genomes.

The term "dark proteome" coined by Perdigão and colleagues, defines regions of proteins that have no detectable sequence homology to other proteins of known three-dimensional structure and therefore cannot be modeled by homology. For 546,000 Swiss-Prot proteins, 44–54% of the proteome in eukaryotes and viruses was found to be "dark", compared with only ~14% in archaea and bacteria.

Proteome Study through Numerous Methods are available to study proteins, sets of proteins, or the whole proteome. In fact, proteins are often studied indirectly, e.g. using computational methods and analyses of genomes. Only a few examples are given below.

Separation Techniques and Electrophoresis

Proteomics, the study of the proteome, has largely been practiced through the separation of proteins by two dimensional gel electrophoresis. In the first dimension, the proteins are separated by isoelectric focusing, which resolves proteins on the basis of charge. In the second dimension, proteins are separated by molecular weight using SDS-PAGE. The gel is dyed with Coomassie Brilliant Blue or silver to visualize the proteins. Spots on the gel are proteins that have migrated to specific locations.

Mass Spectrometry

Mass spectrometry has augmented proteomics. Peptide mass fingerprinting identifies a protein by cleaving it into short peptides and then deduces the protein's identity by matching the observed peptide masses against a sequence database. Tandem mass spectrometry, on the other hand, can get sequence information from individual peptides by isolating them, colliding them with a non-reactive gas, and then cataloguing the fragment ions produced.

In May 2014, a draft map of the human proteome was published in Nature. This map was generated using high-resolution Fourier-transform mass spectrometry. This study profiled 30 histologically normal human samples resulting in the identification of proteins coded by 17,294 genes. This accounts for around 84% of the total annotated protein-coding genes.

Protein Complementation Assays and Interaction Screens

Protein fragment complementation assays are often used to detect protein–protein interactions. The yeast two-hybrid assay is the most popular of them but there are numerous variations, both used in vitro and in vivo.

Proteome and Genome

Genes are the blueprints for proteins. Roughly speaking, genes (DNA) make RNA, and RNA makes proteins.

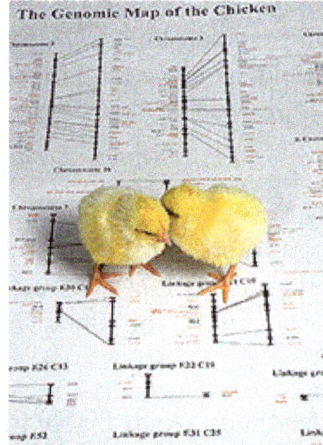

The genomic map of the chicken.

The Human Genome Project mapped the human genome, and now genomes for many other species have been mapped. To a first approximation, each gene makes one protein. If you know the genome for a species, then, you know all its proteins. That is, you can predict the amino acid sequence of each protein. Only about one-third of the human proteins predicted by genomics have been isolated and studied. proteomics is not limited to these well-studied proteins, but also deals with the predicted but as of yet unstudied proteins.

Proteome Over Time

The genome of an organism, its DNA, never changes. In contrast, the proteome changes all the time. The amount of each individual protein varies constantly. Protein levels present problems similar to the word problems of grade school arithmetic. Imagine two bathtubs that are filling and draining at the same time.

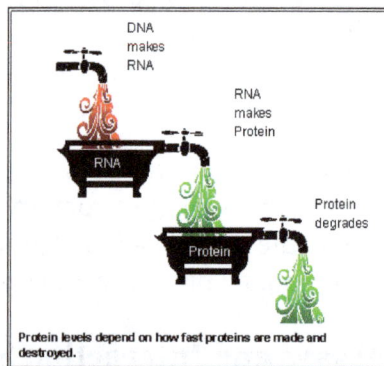

The cascade of DNA to RNA to proteins.

The upper tub is RNA. It fills when the gene is turned on so that its DNA makes RNA. It drains as the RNA makes proteins. The lower tub is the protein, which is filled when the RNA makes protein. This tub is drained as the protein degrades. The amount of protein therefore depends on how open each of the three faucets is. To make things more interesting, within living cells these faucets are continuously and independently adjusted.

Also, as the diagram shows, neither the existence of a gene nor the measured levels of RNA corresponds to measured levels of the protein it makes. The gene is turned on at time zero, and RNA builds up. Later, the protein level starts to rise. It may stay high long after the RNA levels fall.

Proteome and Proteins

How many proteins are in the proteome? Because the human genome has about 20,000 genes, you might guess that the human proteome has about 20,000 proteins. But this number is both too big and too small.

Why too small? The classic dogma of genetics is: One gene equals one protein. This is oversimplified; in fact, there are more proteins than genes.

- Genes are made in sections called exons, which can be spliced together in different ways. About half of all human genes have an alternative splicing, with each splice variant corresponding to a different protein.

- Proteins are modified after they are made. Frequently, the proto-protein made from the gene is cut into sections to form many proteins with different properties. For example, the gene CO3_HUMAN can be spliced to form the proteins C3 beta, anaphylatoxin, C3c, C3d, C3f, and C3g — each with different properties and functions.

Why too big? We usually consider the proteome at the level of only one specific organ or body fluid. A human brain cell and a human liver cell each have the same genome of 20,000 genes, but they have very different proteomes, each with only a subset of all possible human proteins.

- Each tissue or bodily fluid has its own proteome. The human body has many proteomes: the blood, liver, brain, and so on. Furthermore, each organ has many types of cells: the brain has various types of neurons and glial cells as well as blood vessels and blood cells.

- Each cell type has its own proteome. Each of these proteomes have only a small fraction of all the possible proteins.

Plasma Proteome

Plasma is blood without the red and white blood cells. One of the best studied proteomes is that of human blood plasma. The plasma proteome has about 1,000 proteins, not 20,000. What's more, almost all the protein by weight is in the most abundant twenty proteins. All other proteins are present in only trace amounts, roughly a billion times less abundant than the common ones. Also, some proteins appear in several splice forms.

Cellular proteomes have more proteins than plasma, but share these features:

- Not every protein is in each proteome.

- Many proteins in the proteome have several forms.

- A small number of proteins make up the bulk of the proteome.

Proteomics

Proteomics is the large-scale study of proteomes. A proteome is a set of proteins produced in an organism, system, or biological context. We may refer to, for instance, the proteome of a species (for example, Homo sapiens) or an organ (for example, the liver). The proteome is not constant; it differs from cell to cell and changes over time. To some degree, the proteome reflects the underlying transcriptome. However, protein activity (often assessed by the reaction rate of the processes in which the protein is involved) is also modulated by many factors in addition to the expression level of the relevant gene.

Proteomics is used to investigate:

- When and where proteins are expressed,

- Rates of protein production, degradation, and steady-state abundance,

- How proteins are modified (for example, post-translational modifications (PTMs) such as phosphorylation),

- The movement of proteins between subcellular compartments,

- The involvement of proteins in metabolic pathways,

- How proteins interact with one another.

Proteomics can provide significant biological information for many biological problems, such as:

- Which proteins interact with a particular protein of interest (for example, the tumour suppressor protein p53)? Which proteins are localised to a subcellular compartment (for example, the mitochondrion)?

- Which proteins are involved in a biological process (for example, circadian rhythm)?

Several high-throughput technologies have been developed to investigate proteomes in depth. The most commonly applied are mass spectrometry (MS)-based techniques such as Tandem-MS and gel-based techniques such as differential in-gel electrophoresis (DIGE).

Steps in Proteomic Analysis

The following steps are involved in analysis of proteome of an organism as shown in figure:

- Purification of proteins: This step involves extraction of protein samples from whole cell, tissue or sub cellular organelles followed by purification using density gradient centrifugation, chromatographic techniques (exclusion, affinity etc.).

- Separation of proteins: 2D gel electrophoresis is applied for separation of proteins on the basis of their isoelectric points in one dimension and molecular weight on the other. Spots are detected using fluorescent dyes or radioactive probes.

- Identification of proteins: The separated protein spots on gel are excised and digested in gel by a protease (e.g. trypsin). The eluted peptides are identified using mass spectrometry.

Analysis of protein molecules is usually carried out by MALDI-TOF (Matrix Assisted Laser Desorption Ionization-Time of Flight) based peptide mass fingerprinting.

Determined amino acid sequence is finally compared with available database to validate the proteins.

Several online tools are available for proteomic analysis such as Mascot, Aldente, Popitam, Quickmod, Peptide cutter etc.

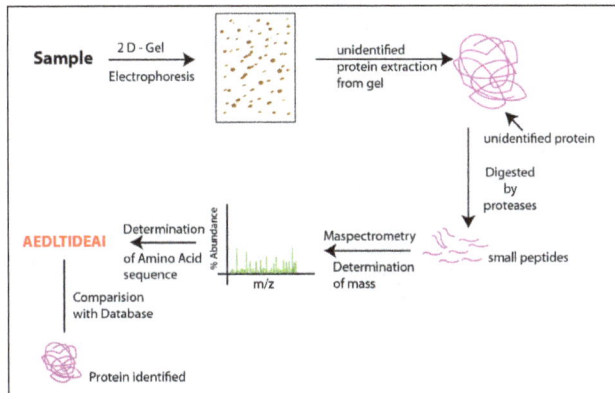

Overview of steps involved in proteomic analysis.

Applications of Proteomics

Proteomics has broad applications in all the aspects of life sciences including several practical applications as drug development against several diseases. Difference in expression protein expression profile of normal and diseased person may be analyzed for target protein. Protein to gene may be predicted. Once protein/gene is identified, function may be predicted. This can help in disease management/drug development.

From Protein to Gene

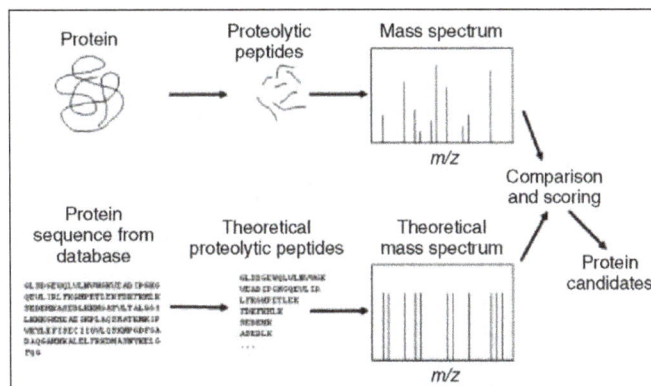

Proteomic analysis-protein to gene sequence.

Whole Genome Sequences of several organisms have been completed but genomic data does not show how proteins function or how these proteins are involved in biological processes. Gene codes for a protein by at several occasion proteins are modified after synthesis (several types of post-translational modifications) for functional diversification.

Proteomics Technologies and their Applications

Proteomics is crucial for early disease diagnosis, prognosis and to monitor the disease development. Furthermore, it also has a vital role in drug development as target molecules. Proteomics is the characterization of proteome, including expression, structure, functions, interactions and modifications of proteins at any stage. The proteome also fluctuates from time to time, cell to cell and in response to external stimuli. Proteomics in eukaryotic cells is complex due to post-translational modifications, which arise at different sites by numerous ways.

Proteomics is one of the most significant methodology to comprehend the gene function although, it is much more complex compared with genomic. Fluctuations in gene expression level can be determined by analysis of transcriptome or proteome to discriminate between two biological states of the cell. Microarray chips have been developed for large-scale analysis of whole transcriptome. However, increase synthesis of mRNA cannot measure directly by microarray. Proteins are effectors of biological function and their levels are not only dependent on corresponding mRNA levels but also on host translational control and regulation. Thus, the proteomics would be considered as the most relevant data set to characterize a biological system.

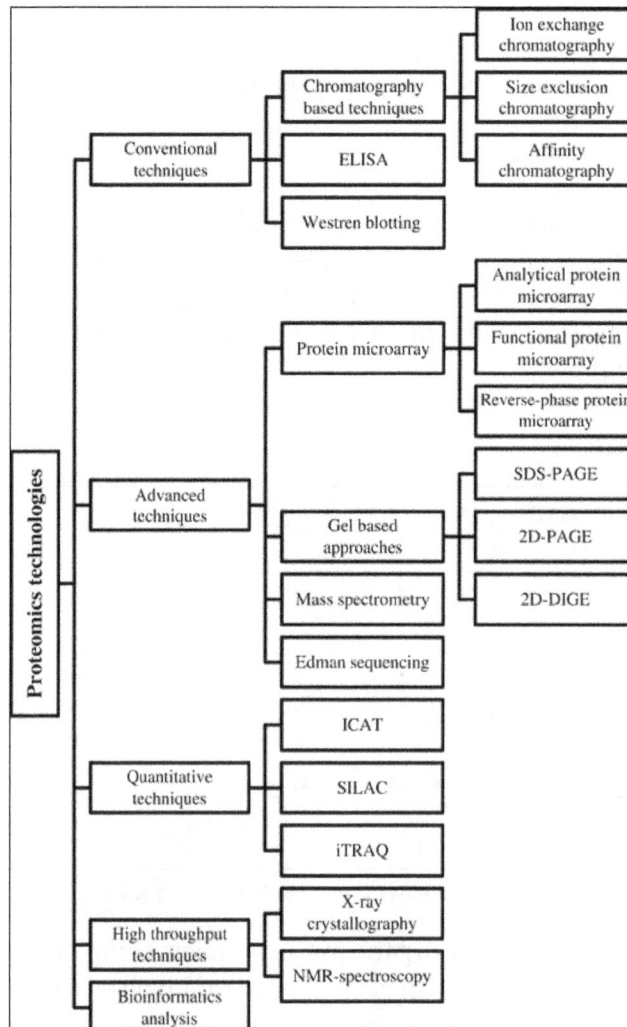

An overview of proteomics techniques.

The conventional techniques for purification of proteins are chromatography based such as ion exchange chromatography (IEC), size exclusion chromatography (SEC) and affinity chromatography For analysis of selective proteins, enzyme-linked immunosorbent assay (ELISA) and western blotting can be used. These techniques may be restricted to analysis of few individual proteins but also incapable to define protein expression level Sodium dodecyl sulfate-polyacrylamide gel electrophoresis (SDS-PAGE), two-dimensional gel electrophoresis (2-DE) and two-dimensional differential gel electrophoresis (2D-DIGE) techniques are used for separation of complex protein samples.

Protein microarrays or chips have been established for high-throughput and rapid expression analysis; however, progress of a protein microarray enough to explore the function of a complete genome is challenging. The diverse proteomics approaches such as mass spectrometry (MS) have developed to analyze the complex protein mixtures with higher sensitivity. Additionally, Edman degradation has been developed to determine the amino-acid sequence of a particular protein. Isotope-coded affinity tag (ICAT) labeling, stable isotope labeling with amino acids in cell culture (SILAC) and isobaric tag for relative and absolute quantitation (iTRAQ) techniques have recently developed for quantitative proteomic. X-ray crystallography and nuclear magnetic resonance (NMR) spectroscopy are two major high-throughput techniques that provide three-dimensional (3D) structure of protein that might be helpful to understand its biological function.

With the support of high-throughput technologies, a huge volume of proteomics data is collected. Bioinformatics databases are established to handle enormous quantity of data and its storage. Various bioinformatics tools are developed for 3D structure prediction, protein domain and motif analysis, rapid analysis of protein–protein interaction and data analysis of MS. The alignment tools are helpful for sequence and structure alignment to discover the evolutionary relationship. Proteome analysis provides the complete depiction of structural and functional information of cell as well as the response mechanism of cell against various types of stress and drugs using single or multiple proteomics techniques. Therefore, this review will emphasized on current progress in proteomics techniques and their applications.

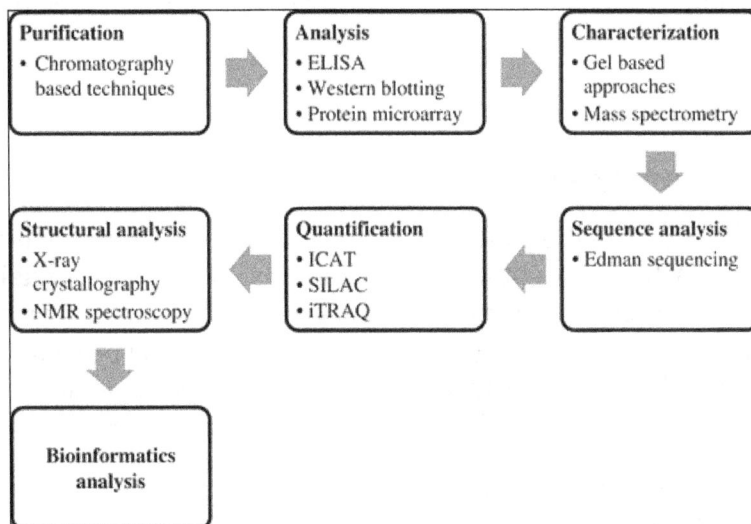

Applications of proteomics techniques.

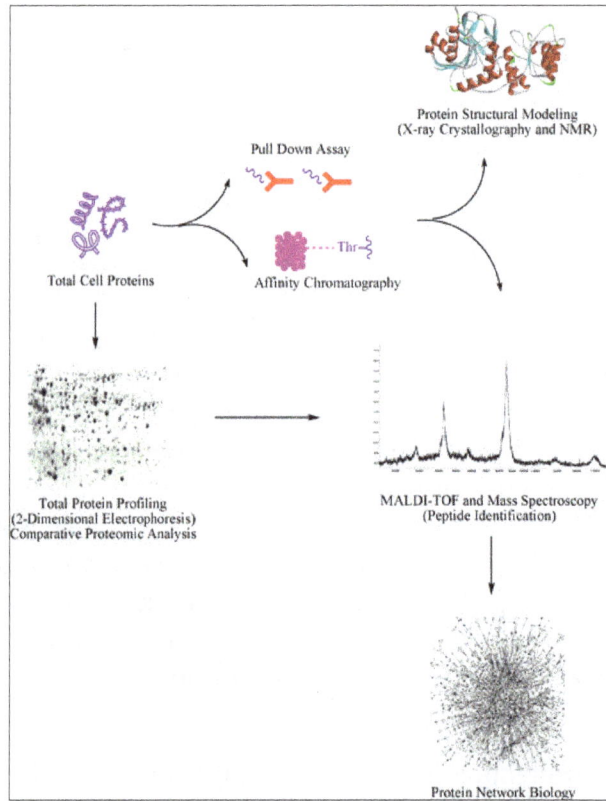

Schematic representation of protein analysis.

The conventional methodology for protein analysis includes protein extraction, purification and structural studies. Cells or tissue are processed by various physical (sonication) and chemical (detergents) techniques for the extraction of total protein. Based upon physiochemical nature of polypeptides, the protein of interest can be separated out by different chromatographic techniques. Various methods including X-ray crystallography, NMR and MALDI-TOF are extensively used for structural elucidation and functional characterization of proteins. Nowadays, high-throughput techniques including total proteome analysis and MALDI-TOF are employed to study protein network biology.

Conventional Techniques

Ion Exchange Chromatography

The IEC is a versatile tool for the purification of proteins on the basis of charged groups on its surface. The proteins vary from each other in their amino-acid sequence; certain amino acids are anionic while others are cationic. The net charged contain by a protein at physiological pH is evaluated by equilibrium between these charges. Initially, it separates the protein on the basis of their charge nature (anionic and cationic), further on the basis of comparative charge strength. The IEC is highly valuable due to its low cost and its capacity to persist in buffer conditions.

A most important virulence factor of Helicobacter pylori is the Neutrophil Activator Protein (HP-NAP) that is able to activate human neutrophil by secreting mediators and reactive oxygen species. The HP-NAP is a potential diagnostic marker for H. pylori and as well a probable drug target and vaccine candidate. One step anionic exchange chromatography has been designated by Shih et al. to purify the recombinant HP-NAP expressed in B. subtilis with 91% recovery. The mussel adhesive proteins (MAPs) have distinctive biocompatible and adhesive properties that are useful for biomedical and tissue engineering. The recombinant MAPs in E. coli and successfully purified through IEC. Antifungal proteins from B. subtilis strain B29 were purified through IEC on diethyl-aminoethyl.

Nigella sativa proteins that retain immune modulatory action have been fractionated through IEC and four peaks were received in complete fractionation. Proteins expressed in transgenic plants commercially values in pharmaceutical products. An example is Aprotinin; an inhibitor of serine proteases that were expressed in corn seed and purified. Cysteine proteases are the key mediators of mammalian apoptosis and inflammation that are expressed in E. coli and purified by Garica-calvo et al. for better understanding of catalytic properties. The serum consists of various chemokines, cytokines, peptide hormones and proteolytic fragments of large proteins that can be purified using strong cation exchange chromatography.

Size Exclusion Chromatography

SEC separates the proteins through a porous carrier matrix with distinct pore size on the basis of permeation; therefore, the proteins are separated on the basis of molecular size. The SEC is robust technique capable of handling proteins in diverse physiological conditions in the presence of detergents, ions and co-factors or at various temperatures. The SEC is used to separate low molecular weight proteins and is a powerful tool for purification of non-covalent multi-meric protein complexes under biological conditions.

The soluble factors produced by Trichomonas vaginalis have the ability to damage the target cells and involved in pathogenesis of trichomoniasis. The phospholipase A2-like lytic factor has been purified and further characterization exhibited 168 and 144 kDa two fractions. The antimicrobial peptides synthesized by marine bacterium Pseodoalteromonas have been purified from culture supernatant through SEC that possess strong inhibitory effect against pathogens involved in skin infections. Cytosolic proteins of Arabidopsis thaliana have been purified to understand how cell coordinates diverse mechanical, metabolic and developmental activities. Purification of intrinsically disordered proteins of A. thaliana was also carried out through SEC. These are expressed during advanced stage of seed development and have a significant role in transcription regulation and signal transduction.

Affinity Chromatography

The affinity chromatography was a major breakthrough in protein purification that enables the researcher to explore protein degradation, post-translational modifications and protein–protein interaction. The basic principle behind the affinity chromatography is the reversible interaction between the affinity ligand of chromatographic matrix and the proteins to be purified.

The affinity chromatography has a wide range of applications in identification of microbial enzymes principally involved in the pathogenesis. Homodimer and heterodimer of HIV-I reverse transcriptase were rapidly purified by metal chelate affinity chromatography. The practical applications of bacteriophages in field of biotechnology and medicine persuade excessive requirement of the phage purification. The T4 bacteriophages have been purified from bacterial debris and other contaminating bacteriophages. The bacterial cells in 'competitive phage display' produced both fusion protein and wild-type proteins. The fusion proteins were integrated into phage capsid and permitted the effective purification of T4 bacteriophages.

A group of amyloid binding proteins interact with different forms of amyloidogenic protein and peptides, therefore modify their pathological and physical role. Affinity chromatography is potentially applied for the diagnosis of Alzheimer's disease by purification of Alzheimer's amyloid peptide from human plasma. The immobilized metal ion affinity chromatography purified the heterologous proteins comprising zinc finger domains. Hexa-histidine affinity tags displayed different affinities to the immobilized metal ions even though both contain same type of domain. However, zinc finger proteins vary in biochemical properties.

Plasma proteins such as factor IX, factor XI, factor VIII, antithrombin III and protein C have been purified through affinity chromatography at industrial scale for therapeutic use. Various ligands have been purified and applied in purification of antibodies. The examples include the lectins for IgM and IgA purification whereas proteins A and G for the purification of IgG molecules.

Enzyme-linked Immunosorbent Assay

In 1971, Engvall and Pearlmann published the first paper on ELISA and quantified the IgG in rabbit serum using the enzyme alkaline phosphatase. The ELISA is highly sensitive immunoassay and widely used for diagnostic purpose. The assay utilizes the antigen or antibodies on the solid surface and addition of enzyme-conjugated antibodies to and measure the fluctuations in enzyme activities that are proportional to antibody and antigen concentration in the biological specimen.

The diagnosis of paratuberculosis or John's disease was made possible by Ethanol Vortex ELISA. The assay distinguished the surface antigens of Mycobacterium avium subspecies paratuberculosis. Capture ELISA was established for detection of Echinostoma caproni in experimentally infected rats. This assay was based on recognition of excretory–secretory antigens by polyclonal rabbit antibodies. The detection limit was 60 ng/ml in fecal sample and 3 ng/ml in sample buffer. Deoxynivalenol (DON), a powerful mycotoxin produced by Fusarium graminearum is a major contaminant of barley and wheat and leads to Fusarium Head Blight. Indirect competitive ELISA for the identification of DON in wheat was developed with detection limit between 0.01 and 100 µg/mL in grains.

Wheat proteins causes allergic reactions in susceptible individuals that have been traced in foods to protect wheat-sensitive individuals using commercially available ELISA kits. Sandwich ELISA was used for the detection of Cry1Ac protein of Bacillus thuringiensis from transgenic BT cotton as their release adversely affect the environment. Indirect competitive ELISA was developed to detect Botrytis cinerea in tissues of fruits. B. cinerea is a phyopathogenic fungus responsible for gray mold and often present as latent infection and deteriorate the healthy fruits. Digital ELISA is capable of detecting single molecule in the blood. The assay was able to detect prostate-specific antigen (PSA) in the serum at low concentration of 14 fg/ml. This assay was capable to detect 1,1-Dichloro-2,2-bis (p-chlorophenyl) ethylene (p,p'-DDE); a metabolite of insecticide and persistent organic pollutant that accumulates in food chain and environment.

Western Blotting

Western blotting is an important and powerful technique for detection of low abundance proteins that involve the separation of proteins using electrophoresis, transfer onto nitrocellulose membrane and the precise detection of a target protein by enzyme-conjugated antibodies. Western blotting is a dominant tool for antigen detection from various microorganisms and is quite helpful in diagnosis of infectious diseases. The seroprevalence of Herpes Simplex Virus type 2 (HSV-2) in African countries was investigated by measuring the specific immunoglobulin G in the sera of patients. Leishmania donovani is responsible for visceral leishmaniasis, which is classically diagnosed by the presence of Hsp83 and Hsp70 antigens in the bone marrow, spleen and liver.

Western blotting was carried out by Li et al. for identification and validation of 10 rice reference proteins. Elongation factor 1-α and heat-shock proteins were the most expressed proteins in rice. Kollerova et al. identified the Plum Pox Virus (PPV) capsid proteins from infected Nicotiana benthamiana. The expression of PfCP-2.9 gene of Plasmodium falciparum in tomato was confirmed through western blot analysis. Specific IgE against Ara h1, Ara h2 and Ara h3 was determined in peanut allergic patients through western blotting.

Edman Sequencing

Edman sequencing was developed by Pehr Edman in 1950 to determine the amino-acid sequence in peptides or proteins. The method comprises chemical reactions that eliminate and identify amino acids residue that is present at the N-terminus of polypeptide chain. Edman sequencing played a major role in development of therapeutic proteins and quality assurance of biopharmaceuticals.

Brucella suis survive and replicate in macrophage due to the acidification. The proteins that are involved in this acidification were identified. Edman degradation and comparison of 13 N-terminal amino-acid sequences revealed that these were signal peptides for its periplasmic location. The protein in B. suis that was involved in membrane permeability at acidic environment was Omp25. The causative agent of hemorrhagic fever, Lassa virus belongs to family of Arenaviridae. The Lassa virus synthesis glycoproteins which are cleaved into GP-1 (amino-terminal subunit) and GP-2 (Carboxy-terminal subunit) after translation and are primarily involved in pathogenesis. The Edman degradation analysis of GP-2 revealed N-terminal tripeptide GTF262.

The prevalence of sesame seed allergy has been increasing due to the use of bakery products and fast-food. The major allergic proteins of Sesamum indicum have been identified from allergic patients through 2D-PAGE and SDS-PAGE and then further analyzed through Edman sequencing. IgE binding epitopes of these proteins were identified that might be helpful in immunotherapeutic approaches. The proteins from leaf sheaths of rice were extracted and analyzed through MS and Edman sequencing to determine its function. The amino-acid sequence of majority of proteins analyzed by both techniques have similar results, therefore suggesting the use of these techniques in combination for the identification of plant protein.

Advanced Techniques

Protein Microarray

Protein microarrays also known as protein chips are the emerging class of proteomics techniques capable of high-throughput detection from small amount of sample. Protein microarrays can be classified into three categories; analytical protein microarray, functional protein microarray and reverse-phase protein microarray.

Analytical Protein Microarray

Antibody microarray is the most representative class of analytical protein microarray. After antibody capture, proteins are detected by direct protein labeling. These are typically used to measure the expression level and binding affinities of proteins. High-throughput proteome analysis of cancer cells was carried out through antibody microarray for differential protein expression in tissues derived from squamous carcinoma cells of oral cavity. Antibody array was also used for protein profiling of bladder cancer. Microarray immunoassay was used for detection of Staphylococcal enterotoxin B, cholera toxin, Bacillus globigii and B. ricin. Analytical and experimental approaches have been developed for identification of cellular signaling pathways and to characterize the plant kinases through protein microarray. Mitogen-activated protein kinases (MAPKs) from Arabidopsis have been characterized. MAPKs are highly conserved single transduction and universal molecules in plants that respond to wide range of extracellular stimuli.

Functional Protein Microarray

Functional protein microarray is constructed by means of purified protein, thus permits the study of various interactions including protein–DNA, protein–RNA and protein–protein, protein–drug, protein–lipid, enzyme–substrate relationship. The first use of functional protein microarray was to analyze the substrate specificity of protein kinases in yeast. Functional protein microarray characterized the functions of thousands of proteins. The protein–protein interaction of A. thaliana was studied and Calmodulin-like proteins (CML) and substrates of Calmodulin (CaM) were identified.

Reverse-phase Protein Microarray

Cell lysates obtained from different cell states are arrayed on nitrocellulose slide that are probed with antibodies against target proteins. Afterwards, antibodies are detected with fluorescent, chemiluminescent and colorimetric assays. For protein quantification, reference peptides are printed on slides. These microarrays are used to determine the altered or dysfunction protein indicative of a certain disease. The analysis of hematopoietic stem cell and primary leukemia samples through reverse-phase protein microarray was found to be highly reproducible and reliable for large-scale analysis of phosphorylation state and protein expression in human stem cells and acute myelogenous leukemia cells. Reverse-phase protein microarray approach was evaluated for quantitative analysis of phosphoproteins and other cancer-related proteins in non-small cell lung cancer (NSCLC) cell lines by monitoring the apoptosis, DNA damage, cell-cycle control and signaling pathways.

Gel-based Approaches

Sodium Dodecyl Sulfate-Polyacrylamide Gel Electrophoresis

SDS-PAGE is a high resolving technique for the separation of proteins according to their size, thus facilitates the approximation of molecular weight. Proteins are capable of moving with electric field in a medium having a pH dissimilar from their isoelectric point. Different proteins in mixture migrate with different velocities according to the ratio between its charge and mass. However, addition of sodium dodecyl sulfate denatures the proteins, therefore separate them absolutely according to molecular weight.

The protein profiling of Mycoplasma bovis and Mycoplasma agalactiae through SDS-PAGE has high diagnostic value as these species are difficult to differentiate with routine diagnostic procedures. The outer membrane proteins from E. coli strains in which ability to form K1 antigen is absent were analyzed through SDS-PAGE. It exhibited varied degree of susceptibility to the human serum. Extracellular protein profile of Staphylococcus spp. was also constructed and their characterization was achieved. The antigenic proteins of Streptococcus agalactiae have been characterized to test the immunogenicity of mastitis vaccine.

The cleome spp. are consumed as green vegetables in African countries and highly valuable for the treatment of cough, fever, asthma, rheumatism and many other diseases. The comparative analysis of leaf and seed proteins of cleome spp. was carried out by SDS-PAGE. The profiling of seed and leaf storage proteins of chickpea (Cicer arientinum) was conducted under drought stress and non-stress conditions. The seed storage proteins of Brassica species are also identified to evaluate the genetic divergence in different genotypes. The influence of heat treatment and addition of demineralized whey on the soluble protein composition of the skim milk was investigated. High molecular weight complexes were formed during the addition of demineralized whey as well as heat treatment which was determined by SDS-PAGE. Large-scale production of insulin is helpful for the management of diabetes, therefore different approaches and species have been used for the production of insulin. Elamin et al. purified and characterized the pancreatic insulin from the Camelus dromedaries.

Two-dimensional Gel Electrophoresis

The two-dimensional polyacrylamide gel electrophoreses (2D-PAGE) is an efficient and reliable method for separation of proteins on the basis of their mass and charge. 2D-PAGE is capable of resolving ~5,000 different proteins successively, depending on the size of gel. The proteins are separated by charge in the first dimension while in second dimension separated on the basis of differences between their mass. The 2-DE is successfully applied for the characterization of post-translational modifications, mutant proteins and evaluation of metabolic pathways. Neidhardt and van Bogelen introduced the highly sensitive technique of 2-DE into the bacterial physiology.

The membrane proteins from the cell wall of Listeria innocua and Listeria monocytogenes involved in the host–pathogen interactions were analyzed with 2-DE and 30 different proteins of two strains were identified. This approach was useful for the comparative study of exotoxins and virulence factors released by enterotoxigenic strains of two food-derived Staphylococcus aureus strains. Pseudomonas aeruginosa secrete numerous proteins during different stages of infection as seen in isolates obtained from cystic fibrosis patients. Current improvements in the 2D-PAGE have been used to study the metabolic system of B. subtilis and a PyrR bacterial regulatory protein was characterized.

Large number of proteins were detected during the seed development in Ocotea catharinensis, and profile was constructed by characterizing these proteins during each developmental stage. Protein extraction from grapes is challenging due to the low concentration of proteins, high activity of proteases and high level of interfering compounds such as polyphenols, flavonoids, terpenes, lignans and tannins; however, Marsoni et al. successfully extracted the proteins from grape tissue through 2-DE. Islam et al. also extracted the proteins from mature rice leaves and applied in the proteome analysis.

Two-dimensional Differential Gel Electrophoresis

2D-DIGE utilizes the proteins labeled with CyDye that can be easily visualized by exciting the dye at a specific wavelength. Cell wall proteins (CWPs) of toxic dinoflagellates Alexandrium catenella labeled with Cy3 have been identified through 2D-DIGE. Quantitative analysis of Brucella suis proteins has been carried out under long-term nutrient starvation and ~30 proteins were identified that vary in concentration among bacteria grown at stationary phase in medium with different nutrient levels. About 70% of regulated proteins showed an increase in expression. The proteins are also involved in regulation, adaptation to harsh condition and transportation. The characterization of proteins expressed in rat neurons have been carried to understand the pathogenesis of West Nile virus.

The plasma membrane responds to the biotic and abiotic stress in plants, therefore the characterization of plasma proteins provides new perception about the plant-specific biological functions. Komatsu characterized the plasma membrane proteome of rice and A. thaliana. The role of apoplastic proteins of 10-day-old rice plants in salt stress response was investigated. For differential analysis, soluble apoplastic proteins from rice shoot stem were extracted and compared with untreated and were found to be involved in oxidation-reduction reaction, carbohydrate metabolism and protein degradation and processing. During ovule development of Pinus tabuliformis, female gametophyte cellularization is a vital process regulated by multiple proteins, which were first extracted in anaphase and prophase then separated through 2D-DIGE.

The biological drugs produced during cell culture technology constitute host cell proteins (HCP) as most important group of impurities. The HCP has diverse molecular and immunological properties and should be effectively monitored and removed during downstream processing. 2D-DIGE was used to screen the HCP composition in CHO cell culture and to compare HCP difference between null cell culture and monoclonal antibody producing cells. The quantitative changes in red blood cell membrane proteins in sickle cell disease were analyzed and the contents of 49 gel spots were found altered by 2.5-fold in comparison with normal cells.

The 2-DE remains a method of choice in proteomic research, though certain limitations enervate its potential as a principal separation technique in modern proteomics. Therefore, the state of the art instrumentation and techniques are rapid expanding as a new means of gel-free analytical techniques. The advancement of MS coupled with shotgun proteomics can find newer directions for sensitive and high quantity protein profiling with more accurate quantification. The chemical label-based approaches remained popular in quantitative proteomics, these methods also have certain drawbacks. The quantitative plant proteomics is more challenging due to problems associated with protein extraction, abundance of proteins in some plants tissues and the lack of well-marked genome sequences. The higher resolution power of MS, exact mass measurements, higher scanning rates and precise chromatogram alignment are essential feature for the successful use of MS in proteomics.

Quantitative Techniques

ICAT Labelling

The ICAT is an isotopic labeling method in which chemical labeling reagents are used for quantification of proteins. The ICAT has also expanded the range of proteins that can be analyzed and permits the accurate quantification and sequence identification of proteins from complex mixtures. The ICAT reagents comprise affinity tag for isolation of labeled peptides, isotopically coded linker and reactive group.

Mycobacterium tuberculosis is considered as a most important human pathogen that contain ~4,000 genes. The proteome analysis was carried out using a combination of Liquid Chromatography (LC), Tandem Mass Spectrometry (MS/MS) and ICAT. The combination of techniques offers comprehensive understanding of biological system and provides additional information. The systemic proteome quantification was carried out possible through ICAT during cell cycle of Saccharomyces cerevisiae that supported the cognition of gene functions. The levels of reactive nitrogen species and reactive oxygen species increase in living cells during abiotic and biotic stress.

The reversible oxidation of protein residues may assist as redox sensors and signal transducers for transmission of anti-stress responses. The thiol group on cysteine residue is sensitive to oxidative species and upon oxidation can modulate protein function. ICAT reagents precisely react with thiol group of cysteine residues, therefore the technique coupled with MS is useful to quantify the thiol-containing redox proteins. The tumor-specific proteins were analyzed through ICAT and MS from the aspirated fluid of breast tumor patients at earlier stages. Beta-globin, hemopexin, lipophilin B and vitamin D-binding proteins were overexpressed while Alpha2HS-glycoprotein was under expressed. It seems that ICAT has potent applications to designate appropriate biomarkers for cancer diagnosis.

Stable Isotopic Labeling with Amino Acids in Cell Culture

SILAC is an MS-based approach for quantitative proteomics that depends on metabolic labeling of whole cellular proteome. The proteomes of different cells grown in cell culture are labeled with "light" or "heavy" form of amino acids and differentiated through MS. The SILAC has been developed as an expedient technique to study the regulation of gene expression, cell signaling, post-translational modifications. Additionally, SILAC is a vital technique for secreted pathways and secreted proteins in cell culture.

SILAC was used for quantitative proteome analysis of B. subtilis in two physiological states such as growth during phosphate and succinate starvation. More than 1,500 proteins were identified and quantified in the two tested states. About 75% genes of B. subtilis were expressed in log phase. Moreover, 10 phosphorylation sties were quantified under phosphate starvation while 35 phosphorylation sites under growth on succinate. Highly purified mutant adenovirus deficient in protein V (internal protein component), wild-type adenovirus and recombinant virus were quantified through SILAC. Viral protein composition and abundance were constant in all types of viruses except virus deficient in protein V which also resulted in reduced amount of another viral core protein.

SILAC was used by for quantitative proteome analysis of A. thaliana. Expression of glutathione S-transferase was analyzed in response to abiotic stress due to salicylic acid and consequent proteins were quantified. Salt stress response and protein dynamics in photosynthetic organism Chlamydomonas reinhardtii have been studied to establish the proteome turnover rate and changes in metabolism under salt stress conditions. RuBisCO was found as the most prominent protein in C. reinhardtii.

The intracellular stability of almost 600 proteins from human adenocarcinoma cells have been analyzed through "dynamic SILAC" and the overall protein turnover rate was determined. Tissue regeneration is imperative in many diseases such as lung disease, heart failure and neurodegenerative disorders. The tissue regeneration and protein turnover rate were quantitatively analyzed in zebra fish. Proteome analysis showed that fin, intestine and liver have high regenerative capacity while heart and brain have the lowest. The proteins in tissue regeneration were mainly involved in transport activity and catalytic pathways.

Isobaric Tag for Relative and Absolute Quantitation

iTRAQ is multiplex protein labeling technique for protein quantification based on tandem mass spectrometry. This technique relies on labeling the protein with isobaric tags (8-plex and 4-plex) for relative and absolute quantitation. The technique comprises labeling of the N-terminus and side chain amine groups of proteins, fractionated through liquid chromatography and finally analyzed through MS. It is essential to find the gene regulation to understand the disease mechanism, therefore protein quantitation using iTRAQ is an appropriate method that helps to identify and quantify the protein simultaneously.

iTRAQ has been applied for quantitative analysis of membrane and cellular proteins of Thermobifida fusca grown in the absence and presence of cellulose. About 181 membrane and 783 cytosolic proteins were quantified during cellulosic hydrolysis. The quantified protein in cellulosic medium was involved in pentose phosphate pathway, glycolysis, citric acid cycle, starch, amino acid, fatty acid, purine, pyrimidine and energy metabolism. Consequently, these proteins have a functional role in cell wall synthesis, transcription, translation and replication. The huge amount of oxidative and hydrolytic enzymes is secreted by Phanerochaete chrysosporium that degrade lignin, cellulose and mixture of lignin and cellulose. The secretory proteins were quantified from P. chrysosporium and 117 enzymes were quantified including cellulose hydrolyzing exoglucanases, endoglucanases, cellobiose dehydrogenase and β-glucosidases.

The presence of soluble aluminum ions (Al^{3+}) in soil limits crop growth; however, *Oryza sativa* are highly aluminum tolerant; therefore, quantitative proteome analysis was carried out in response to Al^{3+} in roots of *O. sativa* at early stages. Out of 700 identified proteins, the expression of 106 proteins was different in Al^{3+} tolerant and sensitive cultivars. The role of hydrogen peroxide (H_2O_2) in growth of wheat was identified through iTRAQ-based quantitative approach that showed that the increased concentration of H_2O_2 restrained the growth of roots and seedlings of wheat. Out of 3,425 identified proteins, 44 were newly identified H_2O_2- responsive proteins involved in detoxification/stress, carbohydrate metabolism and single transduction. Several proteins such as superoxide dismutase, intrinsic protein 1 and fasciclin-like arabinogalactan protein could possibly be involved in H_2O_2 tolerance.

iTRAQ was a useful tool for determination of molecular process involved in development and function of natural killer (NK) cells. Membrane bound proteins of NK cells from CD3-depleted adult peripheral blood cells and umbilical cord blood stem cells were quantified. Ontology analysis exhibited that many of these proteins were involved in nucleic acid binding, cell signaling and mitochondrial functions. Protein profiling was carried out in mouse liver regeneration following a partial hepatectomy. A total of 827 identified proteins, 270 were quantified as well. Fabp5, Lactb2 and Adh1 were downregulated among these while Pabpc1, Mat1a, Oat, Hpx and Dnpep were upregulated.

X-ray Crystallography

X-ray crystallography is the most preferred technique for three-dimensional structure determination of proteins. The highly purified crystallized samples are exposed to X-rays and the subsequent diffraction patterns are processed to produce information about the size of the repeating unit that forms the crystal and crystal packing symmetry. X-ray crystallography has an extensive range of applications to study the virus system, protein–nucleic acid complexes and immune complexes. Further, the three-dimensional protein structure provides detailed information about the elucidation of enzyme mechanism, drug designing, site-directed mutagenesis and protein–ligand interaction.

ZipA and FtsZ are the vital components of spatial ring structure that facilitates cell division in E. coli. ZipA is a membrane anchored protein while FtsZ is homologous of eukaryotic tubulin and their interaction is facilitated by C-terminal domains. X-ray crystallography revealed the structure of C-terminal fragment of FtsZ and binding complex of FtsZ-ZipA. The structure of Norwalk virus that causes gastroenteritis in humans was determined through X-ray crystallography, which revealed that viral capsid consists of 180 repeating units of single protein. The two domains; shell (S) domain and protruding (P) domain of capsid protein are connected by flexible hinge. Eight-standard β-sandwich motif was present in Shell (S) domain while structure of Protruding (P) domain was similar to the domain of eukaryotic translation elongation factor. These domains are the key determinants responsible of cell binding and strain specificity.

The movement of phospholipids, glycolipids, steroids and fatty acids between membranes occurs due to non-specific lipid transfer proteins (nsLTPs). The comparative structure of maize nsLTP in complex with numerous ligands revealed variations in the volume of the hydrophobic cavity depending on the size of bound ligands. The microsomal cytochrome P450 3A4 catalyzes the drug–drug interaction in humans that induce or inhibit the enzymes and metabolically clear the clinically used drugs. The protein structure was analyzed through X-ray crystallography that exhibited a large substrate binding cavity capable to oxidize huge substrates such as statins, cyclosporin, macrolide antibiotics and taxanes. The X-ray crystallography revealed the 3D structure of recombinant horseradish peroxidase in complex with benzohydroxamic acid (BHA). The electron density for BHA was detected in active site of peroxidase along with hydrophobic pocket adjacent to aromatic ring of the BHA.

High-throughput Techniques

Mass Spectrometry

MS is used to measure the mass to charge ratio (m/z), therefore helpful to determine the molecular weight of proteins. The overall process comprises three steps. The molecules must be transformed to gas-phase ions in the first step, which poses a challenge for biomolecules in a liquid or solid phase. The second step involves the separation of ions on the basis of m/z values in the presence of electric or magnetic fields in a compartment known as mass analyzer. Finally, the separated ions and the amount of each species with a particular m/z value are measured. Commonly used ionization method comprises matrix-assisted laser desorption ionization (MALDI), surface enhanced laser desorption/Ionization (SELDI) and electrospray ionization (ESI).

In clinical laboratories, bacterial identification depends on conventional techniques. However, identification of slow growing, fastidious and anaerobic bacteria through conventional techniques is expensive, complex and time consuming. Biswas and Rolain used the MALDI-TOF for early pathogenic bacterial identification, which is useful for early disease control. MS has also became an significant tool in virus research at molecular level, and various viruses and viral proteins including intact viruses, mutant viral strains, capsid protein, post-translational modifications were identified. The study of the changes of viral capsid protein during the infection has allowed the researcher to develop new antiviral drugs. Electrospray ionization mass spectrometry (ESI-MS) coupled with PCR and rRNA gene sequencing provided the accurate and rapid identification of medically important filamentous fungi, yeast and Prototheca species.

Post-translational modification in plants including protein phosphorylation has been distinguished through MS. Top down Fourier Transform mass spectrometry was used to the characterize chloroplast proteins of A. thaliana. Hydrophobic properties and molecular mass of light harvesting proteins of photosystem-II of 14 different plants species were presented by Zolla et al. ESI-MS was used for profiling of integral membrane proteins and detection of post-translational modifications. The most abundant proteins of tomato (Lycopersicon esculentum) xylem sap after Fusarium oxysporum infection were detected with mass spectrometric sequencing and peptide mass finger printing.

The blood proteins including the IBP2, IBP3, IGF1, IGF2 and A2GL have been proposed as biomarkers for the diagnosis of breast cancer. MS was used to characterize these blood proteins. PSA, human growth hormone and interleukin-12 were also analyzed from human serum. Imaging MALDI mass spectrometry was used for the analysis of whole body tissues. The distribution of drugs and metabolites was detected within whole body tissues following drug administration that was useful to analyze novel therapeutics and provide deeper insight into toxicological and therapeutic process.

NMR Spectroscopy

The NMR is a leading tool for the investigation of molecular structure, folding and behavior of proteins. Structure determination through NMR spectroscopy typically involves various phases, each using a discrete set of extremely specific techniques. The samples are prepared and measurements are made followed by interpretive approaches to confirm the structure. The protein structure is fundamental in several research areas such as structure-based drug design, homology modeling and functional genomics.

The three-dimensional structure of transmembrane domain of outer membrane protein A from E. coli has been determined through heteronuclear NMR in dodecylphosphocholine micelles. The fold of protein consists of 19 kDa (177 amino acids) and the structure comprises larger mobile loops toward extracellular side and an eight-stranded β-barrel linked by tight turns on the periplasmic side. The interaction of iso-1-cytochrom c with cytochrome c peroxidase from yeast was investigated by NMR. Chemical shift was observed for both 1H and 15N nuclei arising from the interface of isotopically enriched 15N cytochrome c with cytochrome c peroxidase.

Plant litter decomposition is essential in nitrogen and carbon cycles for the provision of necessary nutrients to the soil and atmospheric CO_2. 15N- and 13C-labeled plant materials were used to monitor the environmental degradation of wheatgrass and pine residues via HR-MAS NMR spectroscopy. The spectra revealed that condensed and hydrolysable tannin were lost from all plant tissues whereas the aliphatic components (cuticles, waxes) and aromatic (partly lignin) persisted along with a small portion of carbohydrate.

Holmes et al. described the variations between metabolic phenotypic from 4,630 participants belonging to 4 human populations through NMR spectroscopy. Metabolic phenotypes including in the study were the products of interactions between variety of factors such as environmental, dietary, genetic and gut microbial activities. Selective metabolites across populations were associated with blood pressure and urinary metabolites that offer the promising discovery of novel biomarkers.

The NMR can be coupled with various approaches like LC or UHPLC to increase the resolution and sensitivity for high-throughput protein profiling. In addition, the structural information can be generated is compared in relation to the identification of metabolites in complex mixtures. NMR coupled with ultra-high performance liquid chromatography (UHPLC) was developed to characterize the metabolic disturbances in esophageal cancer patients for the identification of possible biomarkers for early diagnosis and prognosis. The study revealed considerable alterations in ketogenesis, glycolysis and tricarboxylic acid cycle and amino acid and lipid metabolism in esophageal cancer patients compared with the controls.

Bioinformatics Analysis

Bioinformatics is an essential component of proteomics; therefore, its implications have been progressively increasing with the advent of high-throughput methods that are dependent on powerful data analysis. This new and emergent field is presenting novel algorithms to manage huge and heterogeneous proteomics data and headway toward the discovery procedure.

Endolysins are class of antibacterial enzymes that are becoming useful tool to control spreading of multi-resistant bacteria. The antibacterial property can be altered or expanded by domain swapping, mutagenesis or gene shuffling. The challenge of designing specific endolysins has been revealed in-silico analysis for protein domains present in prophage and phage endolysins. The combination of domains have been studied and sequence type with domain arrangement and conserved amino acids have been determined through multiple sequence alignment. The presence, number and types of binding domain with in endolysins sequence also have been studied. In-silico analysis approach was used to calculate the distribution of the plant food allergens into protein families and determination of conserved surface essential for IgE cross reactivity. The plant food allergen sequences were categorized into four families that indicate the role of conserved structures and biological activities in stimulating allergic properties.

A blood coagulation enzyme, Human Factor Xa (FXa) catalyzes the activation of prothrombin to thrombin and plays an important role in thrombosis and hemostasis. The imbalance in the activation of enzymes intrudes the hemostasis leading to the blood disorders. The safe and effective anticoagulants may be developed by direct inhibition of FXa without effecting thrombin activity essential for normal hemostasis. A study aided the design of more effective ligands through Discovery Studio. Docking studies and binding confirmations revealed that sulfonamide derivatives were inhibitors of FXa.

The use of Bioinformatics for proteomics has gain significantly affluent during the previous few years. The development of new algorithm for the analysis of higher amount of data with increased specificity and accuracy helps in the identification and quantitation of proteins therefore have made possible to achieve expounded data regarding protein expression. The management of such a high quantity of data is the main problem associated with these kind of analyses. Further, it is still difficult to find the association between proteomic data and the other omics technologies including genomics and metabolomics. The database technology along with new semantic statistical algorithms however are the potent tools that might be useful to overcome these limitations.

For MS, the proteins are extracted from the sample and digested using one or several proteases to produce definite set of peptides. Further steps including enrichment and fractionation can be added at protein or peptide level to decrease the complexity of sample or when the analysis of specific subset of proteins is desired. The obtained peptides are analyzed by liquid chromatography coupled with mass spectrometry (LC–MS). Common approaches include either the analysis of deep coverage of proteome by shotgun MS or quantitative investigation for a definite set of proteins through targeted MS. The resulting spectra provide information regarding the sequence, which is important for the identification of proteins. The obtained data may be displayed in a form of 3-D map with mass-to-charge (m/z) ratio, retention time (RT) and intensities of peptides along with fragmentation spectra. The intensity of mass to charge ratio for a particular peptide is plotted along the RTs to get the chromatographic peak. The area under this curve can be used for quantification of peptides, whereas the proteins are identified by the fragmentation spectra. The proteomic data can be uploaded to the repositories that can also be helpful for searching the database. The largest proteome repositories including PRIDE proteomics identification database, Proteome Commons and PeptideAtlas project provide direct access to most of stored data and are valuable tools for data mining.

The protein pathways are a series of reactions inside the cell that exert a particular biological effect. The proteins that are directly involved in reaction along with those that regulates the pathways are combined in pathway databases; therefore, a number of resources and databases are available for the protein pathways. The KEGG, Ingenuity, Pathway Knowledge Base Reactome and BioCarta are some of the pathway databases that include a comprehensive data regarding metabolism, signaling and interactions. In addition to these comprehensive databases, the specific databases for signal transduction pathways such as GenMAPP or PANTHER have been developed. Moreover, databases such as Netpath have been developed, which involve the pathways active in cancer that are helpful for the identification of proteins relevant for a cancer type These public databases possess higher connectivity that allows novel findings for proteins.

The proteins do not act independently in most of the cases and form transient or stable complexes with other proteins. The protein might be intricate as complexes of variable composition and it is essential to study the protein complexes along with the conditions that result in their formation or dissociation for the complete understanding of a biological system. The databases such as BioGRID, IntAct, MINT and HRPD contain the information with reference to protein interactions in complexes STRING is not only a widely used database for protein interaction data, but it connects to various other resources for literature mining. Furthermore, protein networks can be drawn based on the list of genes provided and the available interactions using STRING database

Phosphoproteomics

Phosphoproteomics is a specific type of proteomics that characterizes proteins with the reversible post-translational modification of phosphorylation. Peptide phosphorylation has a vital role in cellular processes such as cell cycle regulation, signal transduction and protein targeting.

Though important to cellular processes, phosphopeptides are found in lower abundance than non-phosphorylated peptides. This means that techniques for detection and quantification of non-phosphorylated peptides have had to be adapted for phosphoproteomics.

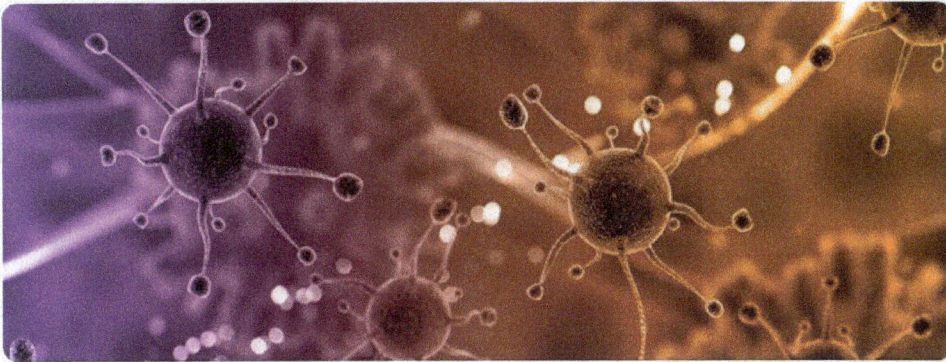

Basic Phosphoproteomic Analysis

The basic methodology for large scale analysis of phosphopeptides first involves SILAC encoding of cultured cells. SILAC, stable isotope labeling by/with amino acids in cell culture, provides differential labeling of cells ready for mass spectrometry (MS). The technique is based on the addition of a 'heavy' and 'light' form of amino acids to proteins. Two cell populations are grown differing only in whether the 'heavy' or 'light' form of amino acid is incorporated. The mass spectrometer differentiates pairs of chemically identical peptides because of the mass difference. Before entering the mass spectrometer, cells are stimulated, lysed and enzymatically digested. The separation of peptides occurs through ion exchange chromatography. A final step of phosphopeptide enrichment then takes place before analysis through mass spectrometry to counteract the low abundance of phosphopeptides.

Different phosphopeptide enrichment techniques are used in phosphoproteomics. In cases where one particular phosphorylated amino acid is being searched for, immunoprecipitation can be used for phosphopeptide enrichment. This involves the employment of antibodies raised against phosphorylated amino acids that bind and isolate the peptide. For large-scale phosphopeptide enrichment, immobilized metal affinity chromatography (IMAC) is commonly used where phosphopeptides are separated according to their affinity for metal ions immobilized on solid resin. Metal oxide affinity chromatography (MOAC) is also employed being composed of a metal oxide or hydroxide matrix, removing the need for resin anchoring. Titanium dioxide is a frequent choice for MOAC because of its known affinity to organic compounds and ability to retain phosphopeptides during high-performance liquid chromatography.

Challenges to Phosphoproteomics

The field of phosphoproteomics is currently limited because of:

1. The low abundance of proteins in comparison to the vast range of cells.

2. The impact of fractional stoichiometry found in phosphorylation.

3. The potential impairment of digestion efficiency during sample preparation.

4. The loss of phosphopeptides during sample preparation.

5. The impairment of phosphopeptide ionization efficiency.

6. The impairment of peptide sequence identification from poor quality MS/MS spectra, caused by the behavior of the labile phosphate group.

7. Problems evaluating the localization of phosphorylation sites.

Phosphopeptide site identification and quantification is more difficult than the measurement of non-phosphorylated proteins. Phosphorylation is a dynamic process prone to errors, meaning that some protein phosphorylation events occur without major functional relevance. When studying large amounts of proteins, random phosphorylation events will be reported because of the high sensitivity of modern mass spectrometers. When random phosphorylation occurs on highly abundant proteins, there is greater difficulty in identifying phosphorylation sites on low abundant proteins. A potential solution to this problem is the formation of temporal profiles for protein phosphorylation upon specific treatment. This would create a strategy for interpreting large amounts of phosphoproteomic data.

Advances in Phosphoproteomics

Advances in MS instrumentation, phosphopeptide enrichment, peptide chromatography and computational proteomics have enhanced the field of phosphoproteomics. Nanoflow liquid chromatography tandem mass spectrometry (LC-MS/MS) in particular is a contemporary method of identifying and quantifying protein phosphorylation. Though SILAC is the most common method of labeling for LC-MS/MS phosphopeptide quantification, alternative methods such as label-free quantification (LFQ) and isobaric tandem mass tags (e.g. iTRAQ, TMT) are increasing in popularity. LFQ works through the comparison of peptide MS signal intensities between MS runs meaning that stable isotope labeling is not required. Isobaric tandem mass tags have an advantage over LFQ which requires the measurement of individual samples; both iTRAQ and TMT can simultaneously measure up to eleven samples in an approach called multiplexing. Choice of quantification techniques is dependent on utilization, with SILAC and LFQ providing greater accuracy in a mixed species comparison with fixed phosphopeptide ratios whilst algorithms for determining phosphorylation sites are improved with multiplexing.

References

- Protein, science: britannica.com, Retrieved 12 April, 2019

- Protein-metabolism, metabolism-and-hormones: diapedia.org, Retrieved 10 March, 2019

- Proteome: nature.com, Retrieved 9 January, 2019

- "UniProt: a hub for protein information". Nucleic Acids Research. 43 (D1): D204–D212. 2014. doi:10.1093/nar/gku989. ISSN 0305-1048. PMC 4384041. PMID 25348405

- Introduction-to-Proteomics: proteomesoftware.com, Retrieved 19 May, 2019

- What-proteomics, proteomics-introduction: ebi.ac.uk, Retrieved 2 February, 2019

- What-is-Phosphoproteomics, life-sciences: news-medical.net, Retrieved 13 April, 2019

Chapter 2

Protein Purification

Protein purification is the sequence of processes that isolate one or a few proteins from a complex mixture. It plays a major role in characterizing the function, structure and interactions of the protein. This chapter discusses in detail the theories related to protein purification along with the different methods used to purify proteins.

Protein separation or purification is a series of processes intended to isolate one or few proteins from a complex mixture. Protein purification is essential for the characterisation of the function, structure and interaction of the protein of interest. Proteins can be separated according to solubility, size, charge and binding affinity.

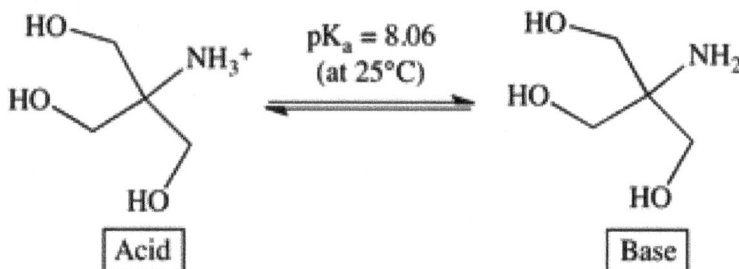

A Tris-buffered solution contains Tris base and its conjugate acid. The pKa of Tris at 25 °C is 8.06, indicating that at pH = 8.06, 50% of the Tris is protonated (in its acidic form) and 50% is deprotonated (in its basic form).

Purified proteins are required for many experimental applications, including structural studies and in vitro biochemical assays. Proteins can be obtained from a tissue or, more often, by their overexpression in a model organism, such as bacteria, yeast, or mammalian cells in culture. Protein purification involves isolating proteins from the source, based on differences in their physical properties. The objective of a protein purification scheme is to retain the largest amount of the functional protein with fewest contaminants. The purification scheme of a protein must be optimized to complete this process in the least number of steps.

Developing a Protein Purification Scheme

The most important consideration in the development of a protein purification scheme is the downstream application of the purified protein. Both the quantity and purity of the protein must be sufficient for experimental analysis. Additionally, information about the behavior of the protein must be taken into consideration, as well-folded and functional protein is required for downstream studies. During purification and subsequent storage, many processes can occur that affect protein quality: protein unfolding, aggregation, degradation, and loss of function. Careful planning to purify protein as quickly as possible and under the most stabilizing conditions will maximize the chance of a successful purification scheme.

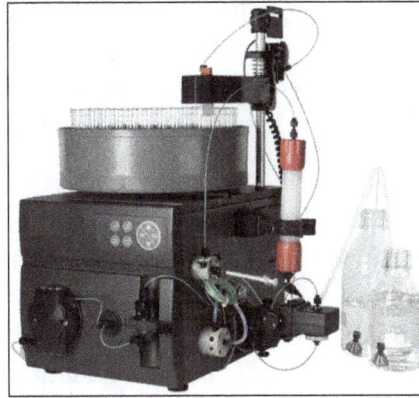

An Aktaprime plus system for the automated chromatographic separation of proteins.

Buffering Component

The solution conditions of a protein at each step of the purification scheme are essential in maintaining protein stability and function. Proteins should be kept in a well-buffered environment to prevent sudden changes in pH that could irreversibly affect their folding, solubility, and function.

A buffer is a solution containing a conjugate acid/base pair. The pH range of a buffer is based on its pK_a, defined as the pH at which 50% of the molecules are in their acidic form, and 50% are in their basic form. A general rule regarding buffers is that the pH of the buffer solution should be within 1.0 pH unit of the pK_a to provide appropriate buffering capacity. This ensures that there is a sufficient amount of the molecule in both its acidic and basic forms to neutralize the solution in case of H^+ or OH^- influx. Thus, buffers prevent pH changes that could negatively affect protein stability.

A good buffer must exhibit the following characteristics:

1. Water solubility.

2. Chemical stability.

3. High buffering capacity at desired pH.

4. Compatibility with analytical and experimental applications.

5. Compatibility with other solution components.

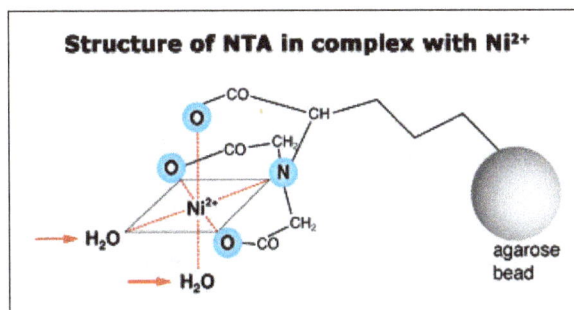

Ni-NTA ligand covalently attached to a cross-linked agarose matrix for the selective purification of polyhistidine-tagged proteins. Histidine residues can coordinate to the Ni^{2+} ion by replacing the bound water molecules (indicated by red arrows). Imidazole or free histidine can then be used to elute the protein, through their ability to coordinate with the Ni^{2+} ion and displace the bound protein.

Many components can serve as biological buffers. The most commonly used buffering components have a near neutral pKa, as they can be used at a physiological pH. Four of the most common biological buffers are listed in Table, along with the pH range at which they can be used, and advantages and disadvantages that might affect their usage in protein purification. Typically, these buffers are used at concentrations above 25mM to ensure adequate buffering capacity.

Buffer	pH range	Advantages and Disadvantages
Phosphate	5.8-8.0	pH is not dependent on temperature Inexpensive. Transparent in the UV range. Cannot be used with divalent cations. Some proteins may precipitate with sodium phosphate buffer, potassium phosphate buffer can be used.
MOPS	6.5-7.9	Cannot be autoclaved . pH is somewhat dependent on temperature . High buffering capacity at physiological pH.
HEPES	6.8-8.2	Cannot be autoclaved. pH is somewhat dependent on temperature. Can form radicals under certain conditions.
Tris	7.5-9.0	pH is dependent on temperature and dilution. Inexpensive. Can interfere with the activity of some enzymes. Transparent in the UV range.

Most commonly used biological buffers for protein purification. Buffers maintain their buffering capacity within a specific pH range, and characteristics of some buffering components could interfere with particular chromatographic procedures or analysis.

Solution Additives

In addition to an appropriate buffering system, solutions used during protein purification from lysis to storage often contain many other components that play a role in facilitating protein purity, stability, and function.

Protease inhibitors are often added to the lysis buffer and in early steps of the purification scheme to prevent degradation of the target protein by endogenous proteases. These are generally not needed toward later stages of the purification, as most or all of the contaminating proteases have been separated from the protein of interest. Metal chelating reagents, such as EDTA or EGTA, are often added to the storage buffer. These metal chelators bind to Mg^{2+} and, thus, prevent cleavage of the purified protein by contaminating metalloproteases. Other additives are often used to protect proteins against damage and enhance their solubility.

Type	Function	Commonly Used Reagents
Reducing Agents	Protect against oxidative damage	2-mercaptoethanol (BME). Dithiothreitol (DTT). Tris (2-carboxyethyl) phosphine (TCEP).
Protease Inhibitors	Inhibit endogenous proteases from degrading proteins	Leupeptin (serine and cysteine protease inhibitor). Pepstatin A (aspartic acid protease inhibitor). PMSF (serine protease inhibitor).
Metal Chelators	Inactivate metalloproteases	EDTA EGTA
Osmolytes	Stabilize protein structure and enhance solubility	Glycerol. Detergents (e.g., CHAPS, NP-40, Triton X-100). Sugars (e.g., glucose, sucrose).
Ionic Stabilizers	Enhance solubility	Salts (e.g., NaCl, KCl, $(NH_4)_2SO_4$.

Additives commonly used in protein purification buffers to increase the stability of proteins.

Additives should only be used if necessary. Trial and error are often required to determine the specific additives that are beneficial to a particular protein purification scheme.

Other Factors to Consider

Other factors also contribute to protein stability during a purification scheme. The least manipulation of a protein during its purification is always the best. Designing a purification scheme that uses the minimum number of steps in the shortest amount of time ensures the highest yield of functional protein. Additionally, it is often best to keep the protein cold throughout the purification. Typically, purification is performed at 4 °C, as this lower temperature both slows down the rate of proteolysis (in the event of contaminating proteases) and promotes structural integrity of proteins.

Impact of Expression System on Protein Purification

Before one can proceed to purify the protein of interest an initial crude sample must be prepared. The first consideration, which takes place well in advance of performing the actual purification, is the source of the protein of interest. This could be a native source such as liver, muscle or brain tissue, though in the post-genomic area it is now relatively rare for investigators to purify proteins from native sources. However, there is still sometimes the need if the investigator wishes to link a catalytic activity to a specific protein sequence.

Nowadays it is far more common for proteins to be purified from recombinant sources. Important decisions need to be made in advance to optimize the subsequent purification. The investigator needs to consider the end-use of the protein (e.g., enzyme assay, structural studies, antibody generation) as this will dictate the quantities and purity required of the final protein preparation.

Which expression system will give the highest level of expression? As a general rule, the higher the expression level, the easier it will be to obtain large quantities of highly purified protein.

If opting for E.coli expression, is the ultimate aim soluble expression or insoluble expression in the form of inclusions bodies? Isolation of inclusion bodies can in itself constitute a significant purification step due to the high abundance of the recombinant protein within inclusion bodies; this has to be balanced against the ease/difficulty of solubilizing and subsequently refolding the target protein from the inclusion bodies and the ultimate yield of the soluble protein. Much work has been done to optimize the yields of functional protein from inclusion bodies.

Another consideration is whether to target intracellular or extracellular (secreted) expression. Intracellular expression requires the protein to be purified away from a significant number of host cellular proteins. By contrast, efficient secretion of the target protein can result in the protein having to be purified away from a small number of secreted host proteins, especially if the host cells are grown under serum-free conditions.

Initial Sample Preparation

Regardless of the source of the target protein, the initial preparation of a crude sample as a starting point for purification is important and needs to be considered at the same time as thinking about the expression and purification strategies.

Extracellular secretion of the target protein may allow for a rapid and straightforward affinity purification protocol to be used; the only sample preparation required may be the decanting of the conditioned media from adherent cells or low-speed centrifugation to remove suspension cells. Of course, it may be necessary to add protease inhibitors or adjust the pH in preparation for the first chromatographic step, but these are simple steps to carry out. However, if ion-exchange is to be performed as the first purification step the sample may first need to be desalted. This can be technically challenging if large volumes of a conditioned medium are to be processed; dialysis or cross-flow filtration are often used depending on the volume of medium to be desalted.

If the target protein is expressed intracellularly, the cells first need to be harvested by centrifugation before being resuspended in an appropriate lysis buffer. As discussed above, the lysis buffer needs to contain an appropriate buffer and other additives to ensure maximum stability of the target protein. The composition of the sample/lysis buffer also needs to be compatible with the subsequent purification step(s) if the investigator is to avoid time-consuming buffer exchange steps before column chromatography.

Next, an efficient method is required for lysing the cells. Various methods have been described in the literature. E.coli cells can be lysed by French press (though this method is not readily scalable), sonication or detergent-based lysis (many commercial lysis reagents are available). The efficiency of detergent-based lysis can be improved by the addition of lysozyme to the lysis buffer. Detergent-based lysis is very mild and does not result in significant shearing of bacterial DNA. Thus the inclusion of DNAase preparations (highly pure preparations are commercially available) is usually required to reduce sample viscosity and produce a sample with good flow characteristics.

Sonication or detergent-based methods can also lyse mammalian and insect cells. DNAase treatment may be required to reduce sample viscosity if there is significant lysis of the nuclear membrane.

Once the cells (microbial/insect/mammalian) have been lysed it is usually necessary to remove cellular debris by centrifugation (typically 15 000 x g for 15 minutes at 4oC to avoid clogging of chromatography columns. The resulting supernatant is the starting point for column chromatographic purification of the target protein.

Need for Protein Purification

Protein purification is either preparative or analytical. Preparative purifications aim to produce a relatively large quantity of purified proteins for subsequent use. Examples include the preparation of commercial products such as enzymes (e.g. lactase), nutritional proteins (e.g. soy protein isolate), and certain biopharmaceuticals (e.g. insulin). Several preparative purifications steps are often deployed to remove bi-products, such as host cell proteins, which poses as a potential threat to the patient's health. Analytical purification produces a relatively small amount of a protein for a variety of research or analytical purposes, including identification, quantification, and studies of the protein's structure, post-translational modifications and function. Pepsin and urease were the first proteins purified to the point that they could be crystallized.

Basic Steps

Extraction

If the protein of interest is not secreted by the organism into the surrounding solution, the first step of each purification process is the disruption of the cells containing the protein. Depending on how fragile the protein is and how stable the cells are, one could, for instance, use one of the following methods: i) repeated freezing and thawing, ii) sonication, iii) homogenization by high pressure (French press), iv) homogenization by grinding (bead mill), and v) permeabilization by detergents (e.g. Triton X-100) and/or enzymes (e.g. lysozyme). Finally, the cell debris can be removed by centrifugation so that the proteins and other soluble compounds remain in the supernatant.

Also proteases are released during cell lysis, which will start digesting the proteins in the solution. If the protein of interest is sensitive to proteolysis, it is recommended to proceed quickly, and to keep the extract cooled, to slow down the digestion. Alternatively, one or more protease inhibitors can be added to the lysis buffer immediately before cell disruption. Sometimes it is also necessary to add DNAse in order to reduce the viscosity of the cell lysate caused by a high DNA content.

Recombinant bacteria can be grown in a flask containing growth media.

Precipitation and Differential Solubilization

In bulk protein purification, a common first step to isolate proteins is precipitation with ammonium sulfate $(NH_4)_2SO_4$. This is performed by adding increasing amounts of ammonium sulfate and collecting the different fractions of precipitate protein. Ammonium sulfate can be removed by dialysis. The hydrophobic groups on the proteins get exposed to the atmosphere, attract other protein hydrophobic groups and get aggregated. Protein precipitated will be large enough to be visible. One advantage of this method is that it can be performed inexpensively with very large volumes.

The first proteins to be purified are water-soluble proteins. Purification of integral membrane proteins requires disruption of the cell membrane in order to isolate any one particular protein from others that are in the same membrane compartment. Sometimes a particular membrane fraction can be isolated first, such as isolating mitochondria from cells before purifying a protein located in a mitochondrial membrane. A detergent such as sodium dodecyl sulfate (SDS) can be used to dissolve cell membranes and keep membrane proteins in solution during purification; however, because SDS causes denaturation, milder detergents such as Triton X-100 or CHAPS can be used to retain the protein's native conformation during complete purification.

Ultracentrifugation

Centrifugation is a process that uses centrifugal force to separate mixtures of particles of varying masses or densities suspended in a liquid. When a vessel (typically a tube or bottle) containing a mixture of proteins or other particulate matter, such as bacterial cells, is rotated at high speeds, the inertia of each particle yields a force in the direction of the particles velocity that is proportional to its mass. The tendency of a given particle to move through the liquid because of this force is offset by the resistance the liquid exerts on the particle. The net effect of "spinning" the sample in a centrifuge is that massive, small, and dense particles move outward faster than less massive particles or particles with more "drag" in the liquid. When suspensions of particles are "spun" in a centrifuge, a "pellet" may form at the bottom of the vessel that is enriched for the most massive particles with low drag in the liquid.

Non-compacted particles remain mostly in the liquid called "supernatant" and can be removed from the vessel thereby separating the supernatant from the pellet. The rate of centrifugation is determined by the angular acceleration applied to the sample, typically measured in comparison to the g. If samples are centrifuged long enough, the particles in the vessel will reach equilibrium wherein the particles accumulate specifically at a point in the vessel where their buoyant density is balanced with centrifugal force. Such an "equilibrium" centrifugation can allow extensive purification of a given particle.

Sucrose gradient centrifugation — a linear concentration gradient of sugar (typically sucrose, glycerol, or a silica based density gradient media, like Percoll) is generated in a tube such that the highest concentration is on the bottom and lowest on top. Percoll is a trademark owned by GE Healthcare companies. A protein sample is then layered on top of the gradient and spun at high speeds in an ultracentrifuge. This causes heavy macromolecules to migrate towards the bottom of the tube faster than lighter material. During centrifugation in the absence of sucrose, as particles move farther and farther from the center of rotation, they experience more and more centrifugal force (the further they move, the faster they move). The problem with this is that the useful separation range of within the vessel is restricted to a small observable window. Spinning a sample twice as long doesn't mean the particle of interest will go twice as far, in fact, it will go significantly further. However, when the proteins are moving through a sucrose gradient, they encounter liquid of increasing density and viscosity. A properly designed sucrose gradient will counteract the increasing centrifugal force so the particles move in close proportion to the time they have been in the centrifugal field. Samples separated by these gradients are referred to as "rate zonal" centrifugations. After separating the protein/particles, the gradient is then fractionated and collected.

Purification Strategies

Chromatographic equipment. Here set up for a size exclusion chromatography.
The buffer is pumped through the column (right) by a computer controlled device.

Choice of a starting material is key to the design of a purification process. In a plant or animal, a particular protein usually isn't distributed homogeneously throughout the body; different organs or tissues have higher or lower concentrations of the protein. Use of only the tissues or organs with the highest concentration decreases the volumes needed to produce a given amount of purified protein. If the protein is present in low abundance, or if it has a high value, scientists may use recombinant DNA technology to develop cells that will produce large quantities of the desired protein (this is known as an expression system). Recombinant expression allows the protein to be tagged, e.g. by a His-tag or Strep-tag to facilitate purification, reducing the number of purification steps required.

An analytical purification generally utilizes three properties to separate proteins. First, proteins may be purified according to their isoelectric points by running them through a pH graded gel or an ion exchange column. Second, proteins can be separated according to their size or molecular weight via size exclusion chromatography or by SDS-PAGE (sodium dodecyl sulfate-polyacrylamide gel electrophoresis) analysis. Proteins are often purified by using 2D-PAGE and are then analysed by peptide mass fingerprinting to establish the protein identity. This is very useful for scientific purposes and the detection limits for protein are nowadays very low and nanogram amounts of protein are sufficient for their analysis. Thirdly, proteins may be separated by polarity/hydrophobicity via high performance liquid chromatography or reversed-phase chromatography.

Usually a protein purification protocol contains one or more chromatographic steps. The basic procedure in chromatography is to flow the solution containing the protein through a column packed with various materials. Different proteins interact differently with the column material, and can thus be separated by the time required to pass the column, or the conditions required to elute the protein from the column. Usually proteins are detected as they are coming off the column by their absorbance at 280 nm. Many different chromatographic methods exist.

Size Exclusion Chromatography

Chromatography can be used to separate protein in solution or denaturing conditions by using porous gels. This technique is known as size exclusion chromatography. The principle is that smaller molecules have to traverse a larger volume in a porous matrix. Consequentially, proteins of a certain range in size will require a variable volume of eluent (solvent) before being collected at the other end of the column of gel.

In the context of protein purification, the eluent is usually pooled in different test tubes. All test tubes containing no measurable trace of the protein to purify are discarded. The remaining solution is thus made of the protein to purify and any other similarly-sized proteins.

Separation based on Charge or Hydrophobicity

Hydrophobic Interaction Chromatography

HIC media is amphiphilic, with both hydrophobic and hydrophilic regions, allowing for separation of proteins based on their surface hydrophobicity. Target proteins and their product aggregate species tend to have different hydrophobic properties and removing them via HIC further purifies the protein of interest. Additionally, the environment used typically employs less harsh denaturing conditions than other chromatography techniques, thus helping to preserve the protein of interest in its native and functional state. In pure water, the interactions between the resin and the hydrophobic regions of protein would be very weak, but this interaction is enhanced by applying a protein sample to HIC resin in high ionic strength buffer. The ionic strength of the buffer is then reduced to elute proteins in order of decreasing hydrophobicity.

Ion Exchange Chromatography

Ion exchange chromatography separates compounds according to the nature and degree of their ionic charge. The column to be used is selected according to its type and strength of charge. Anion exchange resins have a positive charge and are used to retain and separate negatively charged compounds (anions), while cation exchange resins have a negative charge and are used to separate positively charged molecules (cations).

Before the separation begins a buffer is pumped through the column to equilibrate the opposing charged ions. Upon injection of the sample, solute molecules will exchange with the buffer ions as each competes for the binding sites on the resin. The length of retention for each solute depends upon the strength of its charge. The most weakly charged compounds will elute first, followed by those with successively stronger charges. Because of the nature of the separating mechanism, pH, buffer type, buffer concentration, and temperature all play important roles in controlling the separation.

Ion exchange chromatography is a very powerful tool for use in protein purification and is frequently used in both analytical and preparative separations.

Free-flow-electrophoresis

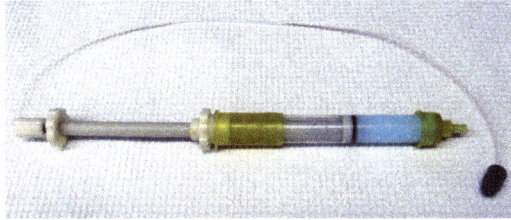

Nickel-affinity column. The resin is blue since it has bound nickel.

Free-flow electrophoresis (FFE) is a carrier-free electrophoresis technique that allows preparative protein separation in a laminar buffer stream by using an orthogonal electric field. By making use of a pH-gradient, that can for example be induced by ampholytes, this technique allows to separate protein isoforms up to a resolution of < 0.02 delta-pI.

Affinity Chromatography

Affinity Chromatography is a separation technique based upon molecular conformation, which frequently utilizes application specific resins. These resins have ligands attached to their surfaces which are specific for the compounds to be separated. Most frequently, these ligands function in a fashion similar to that of antibody-antigen interactions. This "lock and key" fit between the ligand and its target compound makes it highly specific, frequently generating a single peak, while all else in the sample is unretained.

Many membrane proteins are glycoproteins and can be purified by lectin affinity chromatography. Detergent-solubilized proteins can be allowed to bind to a chromatography resin that has been modified to have a covalently attached lectin. Proteins that do not bind to the lectin are washed away and then specifically bound glycoproteins can be eluted by adding a high concentration of a sugar that competes with the bound glycoproteins at the lectin binding site. Some lectins have high affinity binding to oligosaccharides of glycoproteins that is hard to compete with sugars, and bound glycoproteins need to be released by denaturing the lectin.

Metal Binding

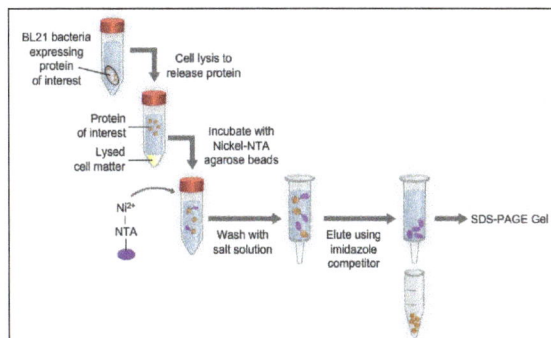

Schematic showing the steps involved in a metal binding strategy for protein purification. The use of nickel immobilized with Nitrilotriacetic acid (NTA) is shown here.

A common technique involves engineering a sequence of 6 to 8 histidines into the N- or C-terminal of the protein. The polyhistidine binds strongly to divalent metal ions such as nickel and cobalt. The protein can be passed through a column containing immobilized nickel ions, which binds the polyhistidine tag. All untagged proteins pass through the column. The protein can be eluted with imidazole, which competes with the polyhistidine tag for binding to the column, or by a decrease in pH (typically to 4.5), which decreases the affinity of the tag for the resin. While this procedure is generally used for the purification of recombinant proteins with an engineered affinity tag (such as a 6xHis tag or Clontech's HAT tag), it can also be used for natural proteins with an inherent affinity for divalent cations.

Immunoaffinity Chromatography

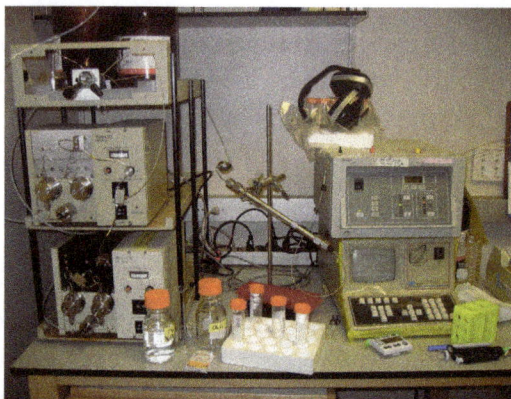

A HPLC. From left to right: A pumping device generating a gradient of two different solvents, a steel enforced column and an apparatus for measuring the absorbance.

Immunoaffinity chromatography uses the specific binding of an antibody-antigen to selectively purify the target protein. The procedure involves immobilizing a protein to a solid substrate (e.g. a porous bead or a membrane), which then selectively binds the target, while everything else flows through. The target protein can be eluted by changing the pH or the salinity. The immobilized ligand can be an antibody (such as Immunoglobulin G) or it can be a protein (such as Protein A). Because this method does not involve engineering in a tag, it can be used for proteins from natural sources.

Purification of a Tagged Protein

Another way to tag proteins is to engineer an antigen peptide tag onto the protein, and then purify the protein on a column or by incubating with a loose resin that is coated with an immobilized antibody. This particular procedure is known as immunoprecipitation. Immunoprecipitation is quite capable of generating an extremely specific interaction which usually results in binding only the desired protein. The purified tagged proteins can then easily be separated from the other proteins in solution and later eluted back into clean solution.

When the tags are not needed anymore, they can be cleaved off by a protease. This often involves engineering a protease cleavage site between the tag and the protein.

HPLC

High performance liquid chromatography or high pressure liquid chromatography is a form of chromatography applying high pressure to drive the solutes through the column faster. This means that the diffusion is limited and the resolution is improved. The most common form is "reversed phase" HPLC, where the column material is hydrophobic. The proteins are eluted by a gradient of increasing amounts of an organic solvent, such as acetonitrile. The proteins elute according to their hydrophobicity. After purification by HPLC the protein is in a solution that only contains volatile compounds, and can easily be lyophilized. HPLC purification frequently results in denaturation of the purified proteins and is thus not applicable to proteins that do not spontaneously refold.

Purified Protein Concentration

A selectively permeable membrane can be mounted in a centrifuge tube. The buffer is forced through the membrane by centrifugation, leaving the protein in the upper chamber.

At the end of a protein purification, the protein often has to be concentrated.

Lyophilization

If the solution doesn't contain any other soluble component than the protein in question the protein can be lyophilized (dried). This is commonly done after an HPLC run. This simply removes all volatile components, leaving the proteins behind.

Ultrafiltration

Ultrafiltration concentrates a protein solution using selective permeable membranes. The function of the membrane is to let the water and small molecules pass through while retaining the protein. The solution is forced against the membrane by mechanical pump, gas pressure, or centrifugation.

Evaluating Purification Yield

The most general method to monitor the purification process is by running a SDS-PAGE of the different steps. This method only gives a rough measure of the amounts of different proteins in the mixture, and it is not able to distinguish between proteins with similar apparent molecular weight.

If the protein has a distinguishing spectroscopic feature or an enzymatic activity, this property can be used to detect and quantify the specific protein, and thus to select the fractions of the separation, that contains the protein. If antibodies against the protein are available then western blotting and ELISA can specifically detect and quantify the amount of desired protein. Some proteins function as receptors and can be detected during purification steps by a ligand binding assay, often using a radioactive ligand.

In order to evaluate the process of multistep purification, the amount of the specific protein has to be compared to the amount of total protein. The latter can be determined by the Bradford total protein assay or by absorbance of light at 280 nm, however some reagents used during the purification process may interfere with the quantification. For example, imidazole (commonly used for purification of polyhistidine-tagged recombinant proteins) is an amino acid analogue and at low concentrations will interfere with the bicinchoninic acid (BCA) assay for total protein quantification. Impurities in low-grade imidazole will also absorb at 280 nm, resulting in an inaccurate reading of protein concentration from UV absorbance.

Another method to be considered is Surface Plasmon Resonance (SPR). SPR can detect binding of label free molecules on the surface of a chip. If the desired protein is an antibody, binding can be translated directly to the activity of the protein. One can express the active concentration of the protein as the percent of the total protein. SPR can be a powerful method for quickly determining protein activity and overall yield. It is a powerful technology that requires an instrument to perform.

Analytical

Denaturing-condition Electrophoresis

Gel electrophoresis is a common laboratory technique that can be used both as preparative and analytical method. The principle of electrophoresis relies on the movement of a charged ion in an electric field. In practice, the proteins are denatured in a solution containing a detergent (SDS). In these conditions, the proteins are unfolded and coated with negatively charged detergent molecules. The proteins in SDS-PAGE are separated on the sole basis of their size.

In analytical methods, the protein migrate as bands based on size. Each band can be detected using stains such as Coomassie blue dye or silver stain. Preparative methods to purify large amounts of protein, require the extraction of the protein from the electrophoretic gel. This extraction may involve excision of the gel containing a band, or eluting the band directly off the gel as it runs off the end of the gel.

In the context of a purification strategy, denaturing condition electrophoresis provides an improved resolution over size exclusion chromatography, but does not scale to large quantity of proteins in a sample as well as the late chromatography columns.

Non-denaturing-condition Electrophoresis

Equipment for preparative gel electrophoresis: Electrophoresis chamber, peristaltic pump, fraction collector, buffer recirculation pump and UV detector (in a refrigerator), power supply and recorder (on a table).

An important non-denaturing electrophoretic procedure for isolating bioactive metalloproteins in complex protein mixtures is quantitative native PAGE. The intactness or the structural integrity of the isolated protein has to be determined by an independent method.

Methods of Protein Purification

Gel Electrophoresis

Gel electrophoresis is a widely known group of techniques used to separate and identify macromolecules as DNA, RNA, or proteins based on size, form, or isoelectric point. The separation of molecules by electrophoresis is based on the fact that charged molecules migrate through a gel matrix upon application of an electric field. These techniques have become a main tool in biochemistry, molecular biology, analytical chemistry and proteomics. Gel electrophoresis is usually used for analytical purposes, but may be a preparative technique to partially purify molecules before applying other techniques, mainly mass spectroscopy to perform proteome analysis.

Although gel electrophoresis is a classical method, in the last decade there has been resurgence in the use of protein electrophoresis with the aim to interpret the great set of data generated by the "omic" techniques. Among them, proteomics may be defined as the comprehensive analysis of the entire protein complement expressed in any biological sample at a given time under specific conditions. The full characterization of the proteome is a formidable challenge as proteins may be subjected to post-translational modifications, have large degrees of dynamic range and be only transiently expressed.

Polyacrylamide Gel Electrophoresis (SDS-PAGE)

Gel electrophoresis of proteins with a polyacrylamide matrix, commonly called polyacrylamide gel electrophoresis (PAGE) is undoubtedly one of the most widely used techniques to characterize complex protein mixtures. It is a convenient, fast and inexpensive method because they require only the order of micrograms quantities of protein. The proteins have a net electrical charge if they are in a medium having a pH different from their isoelectric point and therefore have the ability to move when subjected to an electric field. The migration velocity is proportional to the ratio between the charges of the protein and its mass. The higher charge per unit of mass the faster the migration.

Proteins do not have a predictable structure as nucleic acids, and thus their rates of migration are not similar to each other. They can even not migrate when applying an electromotive force (when they are in their isoelectric point). In these cases, the proteins are denatured by adding a detergent such as sodium dodecyl sulfate (SDS) to separate them exclusively according to molecular weight. This technique was firstly introduced by Shapiro et al. SDS is a reducing agent that breaks disulfide bonds, separating the protein into its sub-units and also gives a net negative charge which allows them to migrate through the gel in direct relation to their size. In addition, denaturation makes them lose their tertiary structure and therefore migration velocity is proportional to the size and not to tertiary structure.

Some highlights of the polyacrylamide gel electrophoresis are:

- Gels suppress the thermal convection caused by application of the electric field, and can also act as a sieving medium, retarding the passage of molecules; gels can also simply serve to maintain the finished separation, so that a post electrophoresis stain can be applied.

- The polyacrylamide gels are formed by polymerization of acrylamide by the action of a cross-linking agent, the bis-acrylamide, in the presence of an initiator and a catalyst. Persulfate ion (S_2O_8-), that is added as ammonium persulfate (APS) is the gel solidifying initiator and a source of free radicals, while TEMED (N, N, N', N'- tetramethylethylenediamine) catalyzes the polymerization reaction by stabilizing these free radicals. In some situations, for example, isoelectric focusing the presence of persulfate can interfere with electrophoresis, so ribofavin and TEMED are used instead.

- Acrylamide solutions are degassed as oxygen is an inhibitor of polymerization. Moreover, the polymerization releases heat that could cause the formation of bubbles within the gel.

- The rate of polymerization is determined by the concentration of persulfate (catalyst) and TEMED (initiator).

- The ratio between of acrylamide/bisacrylamide as well as the total concentration of both components, affects the pore size and rigidity of the final gel matrix. These, in turn, affect the range of protein sizes that can be resolved. The size of the pores created in the gel is inversely related to the amount of acrylamide used. For instance, a 7% polyacrylamide gel has larger pores than a 12% polyacrylamide gel. Gels with a low percentage of acrylamide are typically used to resolve large proteins, and high percentage gels are used to resolve small proteins. "Gradient gels" are specially prepared to have low percent-acrylamide at the top and high percent-acrylamide at the bottom, enabling a broader range of protein sizes to be separated.

The acrylamide gel electrophoresis systems may be performed using one or more buffers, in these cases we speak of continuous phosphate buffer system or discontinuous buffer systems. Laemmli adopted the discontinuous electrophoresis method and the term "Laemmli buffer" is often used to describe the tris-glycine buffer system that is utilized during SDS-PAGE.

In discontinuous systems the first buffer ensures the migration of all proteins in the front of migration, what causes the accumulation of the entire sample that has been loaded into the well. The separation really begins from the moment when the migration front reaches the boundary of the second buffer. The first gel, "stacking", has larger pore (lower percentage of acrylamide/bisacrylamide) and has a pH more acidic than the second gel which is what really separates proteins. This system is particularly suitable for analyzing samples diluted without losing resolution.

The resolution of peptides below 14 kDa is not sufficient in conventional tris-glycine systems. This problem was solved by the development of a new system by Schägger and von Jagow. In this method an additional spacer gel is introduced, the molarity of the buffer is increased and tricine is used as terminated ion instead of glycine. This method yields linear resolution from 100 to 1 kDa.

Detection of proteins in gels Proteins separated on a polyacrylamide gel can be detected by various methods, for instance dyes and silver staining.

Detection of Proteins in Gel

- Dyes: The Coomassie blue staining allows detecting up to 0.2 to 0.6 µg of protein, and is quantitative (linear) up to 15 to 20 µg. It is often used in methanol-acetic acid solutions and is discolored in isopropanol-acetic acid solutions. For staining of 2-DE gels it is recommended to remove ampholytes by adding trichloroacetic (TCA) to the dye and subsequently discolor with acetic acid.

- Silver staining: It is an alternative to routine staining protein gels (as well as nucleic acids and lipopolysaccharides) because its ease use and high sensitivity (50 to 100 times more sensitive than Coomassie blue staining). This staining technique is particularly suitable for two-dimensional gels.

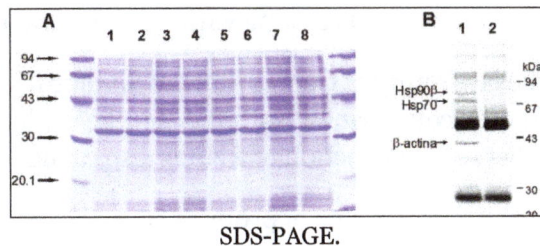

SDS-PAGE.

Proteins separated on SDS-PAGE and detected by Coomassie blue (A) and silver staining (B). Standards of proteins to know molecular weight are also loaded at edges.

- Detection of radioactive proteins by autoradiography: The autoradiography is a detection technique of radioactively labeled molecules that uses photographic emulsions sensitive to radioactive particles or light produced by an intermediate molecule. The emulsion containing silver is sensitive to particulate radiation (alpha, beta) or electromagnetic radiation (gamma, light), so that it precipitates as metallic silver. The emulsion will develop as dark precipitates in the region in which radioactive proteins are detected.

Native Page

Depending on the state of the protein (native or denatured) along the electrophoretic process, the techniques are classified into native and denaturing electrophoresis.

- Denaturing electrophoresis, the most common, is when the protein undergoes migration ensuring complete denaturation (loss of three-dimensional structure). In this case, migration is proportional to the load and the size of the molecule but not to its form. The most commonly used denaturing agent is the detergent SDS.

- Native electrophoresis is when the protein undergoes migration without denaturation. In this situation, proteins migrate according to their charge, size and shape. Furthermore, in some cases the interactions between subunits and between proteins are kept, separating at the level of complexes. Buffer systems used in this electrophoresis are: Tris-glycine (pH range 8.3 to 9.5), tris-borate (pH range 7.0 to 8.5) and tris-acetate (pH range 7.2 to 8.5).

- Blue native electrophoresis permits a high resolution separation of multiprotein complexes under native conditions. This technique consists of polyacrylamide gel electrophoresis where the nondenaturing compound Coomassie blue G-250 is added to both the sample and to the electrophoresis buffers to confer a negative charge on the protein complexes so they can migrate intact toward the anode. Using this methodology, many samples can be concurrently separated during a single electrophoretic run, and a direct comparison of protein complexes readily allows for the identification of differences in protein expression and direct further functional analysis.

Immunodetection of Proteins by Western Blot

Western blot is a widely used method in molecular biology and biochemistry to detect proteins in a sample of cell homogenate or extract. The proteins are transferred from the gel to a membrane -made of nitrocellulose, nylon or polyvinylidene difluoride (PVDF)-, where they are examined using specific antibodies to the protein. As a result, the amount of protein in a sample can be examined and it is also possible to compare levels among various analytical groups.

The method was initiated in the laboratory of George Stark at Stanford. The name "western blot" was given to the technique by Burnette, comparing it with the "southern blot" technique for DNA detection developed by Edwin Southern. The detection of RNA is also called Northern blotting.

The most powerful method is the transference of proteins from the gel to a membrane by applying an electric field perpendicular to the gel. There are however other methods of transferring or applying a protein on the membrane. The simplest is to apply it directly as a small drop of a concentrated solution on the membrane. The absorption of the drop causes the adhesion of the protein to the membrane, leaving it as a spot or "dot" (this is the case of the "dot blot"). There are devices that make possible the application of proteins to the membrane directly, using a suction that facilitates the penetration of the solution, and are named "dot blot" or "slot blot" on the basis that the proteins were applied as a circular drop or a line.

Working with proteins bound to a membrane has advantages over employment within the gel:

- Staining and discolor are faster.

- No staining occurs to ampholytes in isoelectric focusing gels.

- Smaller amounts of proteins are detected as they are concentrated at the surface and not diluted across the thickness of the gel.

- The membranes are much easier to manipulate than the gel itself.

Blotting Procedure

It consists of 5 stages:

1. Immobilization of proteins on the membrane either by transference (electrophoresis, suction, pressure) or by direct application. The procedure starts piling a flat sponge on filter paper soaked in transference buffer, the gel, the membrane in direct contact with the gel plus filter paper and finally a flat sponge. This set is included between two layers of

perforated plastic and placed in a tank which is a saline solution (transference buffer) and two plate electrodes (designed to achieve a uniform field across the surface of the gel). They are disposed so that the gel is toward the anode (-) and the membrane to the cathode (+).

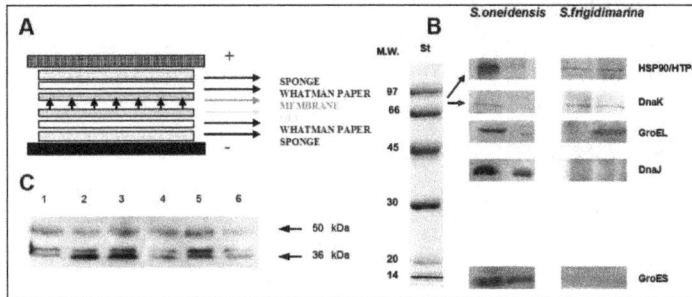

Western blot.

Scheme of the components used for immobilization of proteins on a membrane by western blot (A). Examples of western blot performed in extracts from bacterial cells (B) and rat neurons (C) with specific antibodies.

2. Saturation of all binding sites of proteins in the membrane not occupied to prevent nonspecific binding of antibodies, which are proteins.

3. Incubation with primary antibody against the protein of interest.

4. Incubation with secondary antibodies, or reactives acting as ligands of the primary antibody bound to enzymes or other markers. The proteins labeled with enzymes are visible by incubation with appropriate substrates to form insoluble colored products in the place where the protein were. There are several possibilities:

Enzyme coupled secondary antibodies: an antibody to the specific binding antibody is conjugated to the enzyme peroxidase or alkaline phosphatase.

Another possibility is the use of an amplifying enzyme which is part of a biotin-avidinperoxidase complex or a complex with alkaline phospohatase.

Enhanced chemiluminescence (ECL) is other commonly used method for protein detection in western blots. ECL is based on the emission of light during the horse radish peroxides (HRP)- and hydrogen peroxide-catalyzed oxidation of luminol. The emitted light is captured on film or by a CCD camera, for qualitative or semiquantitative analysis. This method allows stripping and re-probing the blot with different antibodies. Two examples of western blot with different antibodies can be seen in figure above.

Isoelectric Focusing (IEF)

This technique is based on the movement of molecules in a pH gradient. Amphoteric molecules such as amino acids and proteins are separated in an environment where there is a difference of potential and pH gradient. The region of the anode (+) is acidic and the cathode (-) is alkaline. Between them down a pH gradient such that the molecules to be separated have their isoelectric point within the range. Substances that are initially in regions with a pH below its isoelectric point are positively charged and migrate towards the cathode, while those that are in media with pH lower than its pI will have negative charge and migrate towards the anode. The migration will lead to a region where the pH coincide with its pI, have a zero net charge (form zwitterions) and stop. Thus amphoteric molecules are located in narrow bands where the pI coincides with the pH. In this technique the point of application is not critical as molecules will always move to their pI region. The stable pH gradient between the electrodes is achieved using a mixture of low molecular weight ampholytes which pI covers a preset range of pH.

Two-dimensional Gel Electrophoresis

Two-dimensional gel electrophoresis (2-DE) is based on separating a mixture of proteins according to two molecular properties, one in each dimension. The most used is based on a first dimension separation by isoelectric focusing and second dimension according to molecular weight by SDS-PAGE.

The general workflow in a 2-DE experiment would be:

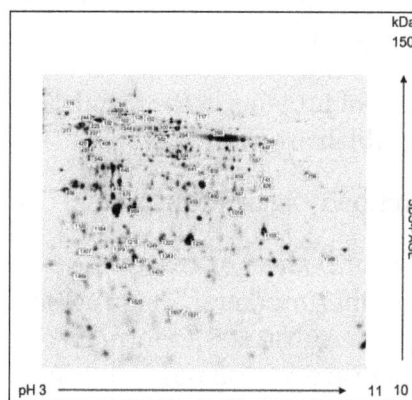

2-DE preparative gels.

Proteins of Chlamydomonas reinhardtii resolved by 2-DE from preparative gels stained with MALDI-MS compatible silver reagent for peptide mass fingerprinting analysis. First dimension: isoelectric focusing in a 3-11 pH gradient. Second dimension: SDS-PAGE in a 12% acrylamide (2.6% crosslinking) gel (1.0 mm thick). Numbered spots marked with circle correspond to proteins compared to be subsequently identified by MALDI-TOF MS.

- Sample Preparation

The method of sample preparation depends on the aim of the research and is crucial to the success of the experiment. Factors such as the solubility, size, charge, and isoelectric point (pI) of the proteins of interest enter into sample preparation. Sample preparation is also important in reducing the complexity of a protein mixture. The protein fraction to be loaded on a 2-DE gel must be in a low ionic strength denaturing buffer that maintains the native charges of proteins and keeps them soluble.

- First-Dimension Separation

This part is performed by IEF. Using this technique, proteins are separated on the basis of their pI, the pH at which a protein carries no net charge and will not migrate in an electrical field.

- Equilibration

A conditioning step is applied to proteins separated by IEF prior to the second-dimension run. This process reduces disulfide bonds and alkylates the resultant sulfhydryl groups of the cysteine residues. Concurrently, proteins are coated with SDS for separation on the basis of molecular weight.

- Second-Dimension Separation

This part is performed by SDS-PAGE. The choice for the gel depends on the protein molecular weight range to be separated. The ability to run many gels at the same time and under the same conditions is important for the purpose of gel-to-gel comparison.

- Staining

In order to visualize proteins in gels, they must be stained in some manner. The selection of staining method is determined by several factors, including desired sensitivity, linear range, ease of use, expense, and the type of imaging equipment available. At present there is no ideal universal stain. Sometimes proteins are detected after transference to a membrane support by western blotting.

- Image Analysis

The ability to collect data in digital form is one of the major factors that enable 2-DE gels to be a practical means of collecting proteome information. It allows unprejudiced comparison of gels and cataloging of immense amounts of data. Many types of imaging devices interface with software designed specifically to collect, interpret, and compare proteomics data. One of the biggest problems in 2-DE is the analysis and comparison of complex mixtures of proteins. Currently there are databases capable of comparing two-dimensional gel patterns. These systems allow automatic comparison of spots for the precise identification of those needed in the quantitative analysis.

- Protein Identification

Once interesting proteins are selected by differential analysis or other criteria, the proteins can be excised from gels, distained and digested to prepare their identification by mass spectrometry (Fig. 5 A). This technique is known as peptide mass fingerprinting. The ability to precisely determine molecular weight by matrix-assisted laser desorption/ionizationtime of flight mass spectrometry (MALDI-TOF MS) and to search databases for peptide mass matches has made high-throughput protein identification possible. Proteins not identified by MALDI- TOF can be identified by sequence tagging or de novo sequencing using the Q-TOF electrospray LC-MS-MS.

Two-dimensional Fluorescence Difference Gel Electrophoresis (2-D DIGE)

2-D Fluorescence Difference Gel Electrophoresis (2-D DIGE) is a method that labels protein samples prior to 2-DE, enabling accurate analysis of differences in protein abundance between samples. It is possible to separate up to three different samples within the same 2-DE gel. The technology is based on the specific properties of fluorescent cyanine dyes that are spectrally resolvable and size- and charge-matched. Identical proteins labeled with each of the three dyes (Cy2, Cy3 and Cy5) will migrate to the same position on a 2-DE gel. This ability to separate more than one sample on a single gel permits the inclusion of up to two samples and an internal standard (internal reference) in every gel. The internal standard is prepared by mixing together equal amounts of each sample in the experiment and including this mixture on each gel.

There are several analysis software programs developed to exploit the advantages of fluorescent dyes. They enable the detection, quantization, matching, and analysis of gels. The algorithm in this type of software co-detects overlaid image pairs and produces identical spot boundaries for each pair. This enables direct spot volume ratio measurements and therefore produces an accurate comparison of every protein with its representative ingel internal standard. The software automatically performs detection, background subtraction, quantization, and normalization, which takes into account any differences in the dyes, i.e. molar extinction coefficient, quantum yields, etc.

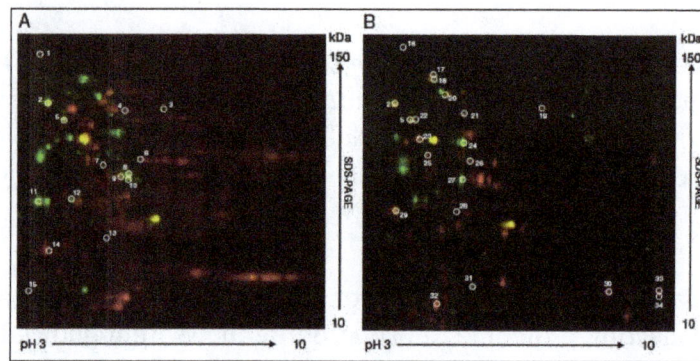

Example of 2D-DIGE gels.

Cell extracts were obtained from S. oneidensis cultured at 30 °C (A) and 4 °C (B), and subjected to immunoprecipitation with monoclonal antibody. Immunoprecipitates were resolved by 2D-DIGE gels following image analysis. Numbered spots marked with circles corresponded to proteins identified by MALDI-TOF MS.

Protein Identification by Matrix-assisted Laser Desorption/Ionization-time of Flight (MALDI-TOF) Mass Spectrometry

Mass spectrometry is a technique to analyze with high accuracy the composition of different chemical elements and atomic isotopes splitting their atomic nuclei according to their masscharge ratio (m/z). It can be used to identify different chemical elements that form a compound or to determine the isotopic content of different elements in the same compound.

Firstly, the material to be analyzed is ionized and ions are then transported by magnetic or electric fields to the mass analyzer. Techniques for ionization have been key to determine what types of samples can be analyzed by mass spectrometry. Two techniques are often used with liquid and solid biological samples: Electro spray ionization and laser matrixassisted laser desorption/ ionization (MALDI). In the MALDI ionization analytes cocrystallized with a suitable matrix are converted into ions by the action of a laser. This source of ionization is usually associated with a time of flight analyzer (TOF) in which the ions are separated according to their mass-charge after being accelerated in an electric field. At last, a mass spectrometer detector records the charge induced or current produced when an ion passes by or hits a surface. A mass spectrum is registered for each protein.

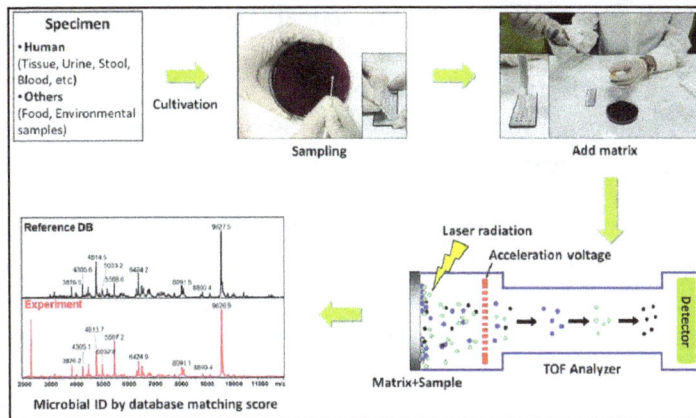

Microbial ID by database matching score

Software and Database Search Algorithms to Analyze Spectral Data

A variety of tools and commercially available software exist that allow for protein identification from peptide sequences determined by mass spectrometry (or other sequencing techniques).

Some examples of database search programs and algorithms are:

- SEQUEST identifies collections of tandem mass spectra to peptide sequences that have been generated from databases of protein sequences. It was one of the first, if not the first, database search program. While very successful in terms of sensitivity, it is quite slow to process data and there are concerns against specificity, especially if multiple posttranslational modifications (PTMs) are present.

- Mascot is a powerful search engine that uses mass spectrometry data to identify proteins from primary sequence databases.

- Scaffold 3 is a software which produces a confidence level for protein identification from one or more Mascot, Sequest, X. Tandem, or Phenyx searches. It can be used in conjunction with a MS/MS search engine in order to validate/visualize data across multiple experimental runs as well as provide a more accurate protein probability.

Chromatographic Methods

To be able to isolate a specific protein from a crude mixture the physical and/or chemical properties of the individual protein must be utilized. There is no single or simple way to purify all kinds of proteins. Procedures and conditions used in the purification process of one protein may result in the inactivation of another. The final goal also has to be considered when choosing purification method. The purity required depends on the purpose for which the protein is needed. For an enzyme that is to be used in a washing powder, a relatively impure sample is sufficient, provided it does not contain any inhibiting activities. However, if the protein is aimed for therapeutic use it must be extremely pure and purification must then be done in several subsequent steps.

The aim of a purification process is not only removal of unwanted contaminants, but also the concentration of the desired protein and the transfer to an environment where it is stable and in a form ready for the intended application.

In the early days of protein chemistry, the only practical way to separate different types of proteins was by taking advantage of their relative solubility. Part of a mixture was caused to precipitate through alteration of some properties of the solvent e.g. addition of salts, organic solvents or polymers, or varying the pH or temperature. Fractional precipitation is still frequently used for separation of gross impurities, membrane proteins and nucleic acids.

Under certain conditions, proteins adsorb to a variety of solid phases, preferably in a selective manner. Calcium phosphate gels have frequently been used to specifically adsorb proteins from heterogeneous mixtures. The adsorption principle is further explored in column chromatography. Due to their high resolving power, different chromatography techniques have become dominant for protein purification.

Chromatography

Chromatography refers to a group of separation techniques that involves a retardation of molecules with respect to the solvent front that progresses through the material. The name literally means "color drawing" and was originally used to describe the separation of natural pigments on filter papers by differential retardation. The same principle is now commonly used for protein separation. Column chromatography is the most common physical configuration, in which the stationary phase is packed into a tube, a column, through which the mobile phase, the eluent, is pumped. The degree to which the molecule adsorbs or interacts with the stationary phase will determine how fast it will be carried by the mobile phase. Chromatographic separation of protein mixtures has become one of the most effective and widely used means of purifying individual proteins.

General properties of proteins are used to isolate them from other, non-protein contaminants. Minor differences between various proteins, such as size, charge, hydrophobicity and biospecific interaction are used to purify one protein from another.

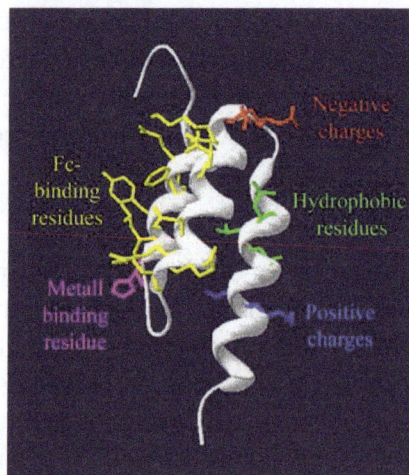

Selective protein properties Examples of properties that are
used to separate one protein from another.

In a typical chromatography process the first step is a capturing step, where the product binds to the adsorbent while the impurities do not. Further, weakly bound proteins are washed away before the conditions are changed so that the target protein is eluted. Several versions of liquid chromatography, differing mainly in the types of stationary phase, are used for protein purification. The process of Size exclusion Chromatography is a bit different since separation is based on sieving properties of the stationary phase and not on adsorption. Two variants of column chromatography, Ion Exchange Chromatography and Size Exclusion Chromatography, are illustrated in figure below.

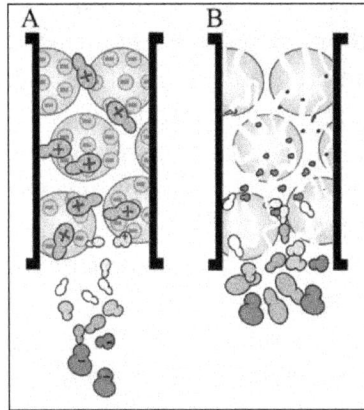

Illustrations of two classical chromatographic methods:
A. Ion Exchange Chromatography. The charges of a protein are used for purification.
B. Size Exclusion Chromatography. Protein size is used for fractionation.

Ion Exchange Chromatography

Ionic interactions are the basis for purification of proteins by Ion Exchange Chromatography (IEXC). The separation is due to competition between proteins with different surface charges for oppositely charged groups on an ion exchanger adsorbent.

The fundamental theory of IEXC has a very long history. One of the first examples of ion exchange purification is attributed to Moses, who purified acrid water with the aid of a special type of wood (2 Mos. 15:25). The first synthetic ion exchangers consisted of hydrophobic polymer matrices, highly substituted with ionic groups. Due to their low permeability these matrices had low capacities for larger molecules such as proteins. In addition, the hydrophobic nature of the matrix denatured the proteins. It was not until hydrophilic materials of macroporous structure were introduced, in the late 1950s, that ion exchange chromatography of biological macromolecules became a useful separation tool.

Charge Properties of Proteins

Proteins are complex ampholytes that have both positive and negative charges. The isoelectric point (pI) of a protein (the pH at which the net charge is zero) depends on the proportions of ionizable amino acid residues in its structure. Positive charges are usually provided by arginines, lysines and histidines, depending of the pH of the surrounding buffer. Any free N-terminal amine will also contribute with a positive charge below pH 8. Negative charges are principally provided by aspartate and glutamate residues and the C-terminal carboxyl group. Virtually all these residues are ionized above pH 6. At higher pH values (>8) cysteines may become ionized too.

The charged groups nearly always reside on the protein surface. Exceptions are mainly metallo-proteins, where an internal metal ion often is coordinated by charged residues. Influences from neighboring groups and the position in the tertiary structure will affect the pKa for the side-chain groups. The combined influence of all of the charged side chains will give the protein a varying net charge depending on the pH of the solute.

Therefore, it is possible to separate proteins using either fixed positive charges on the stationary phase, anion exchanger, or fixed negative charges, cation exchanger. A protein must displace the counterions to become attached and consequently the net charge on the protein will be the same as that of the counterions displaced, thereof the term "ion exchange". Generally, anion exchange chromatography (AIEXC) is carried out at pH values above the isoelectric point of the protein of interest, while cation exchange chromatography (CIEXC) is carried out below the isoelectric point. The pH interval in which ion exchange chromatography is carried out is restricted by the pH range in which the protein is stable. To achieve good adsorption, the pH of the buffer chosen should be at least one pH unit above or below the isoelectric point of the analytes to be separated. The pH in the microenvironment of an ion exchanger is not exactly the same as that of the applied buffer. This is due to the so called Donnan effect, that protons are attracted or expelled (depending on charge of the matrix) from the microenvironment close to the matrix. In general, the pH close to the matrix is up to 1 unit higher than that in the surrounding buffer in anion exchangers, and 1 unit lower in cation exchangers. Consequently, if a protein is adsorbed on a cation exchanger at pH 5, it will be exposed to pH 4, and if it has poor stability at that pH, it may be denatured.

Most proteins are negatively charged at physiological pH values (pH 6 – 8) and therefore, in many applications of IEXC a first approach is to use an anion exchanger. However, for optimal separation the selection of pH should aim to, if possible, introduce as large charge difference as possible between the target protein and the contaminants. A distribution profile of the isoelectric points of a wild-type E. coli proteome showed that 95% of the intracellular proteins were negatively charged at basic pH and should therefore most easily be removed using CIEXC.

It is possible to use IEXC more than once in a purification strategy since the pH of the separation can be modified to alter the charge characteristics of the sample components. Typically IEXC is used to bind the target molecule and then wash away non-bound contaminants. However, the technique can also be used to bind the impurities if required. In that case the protein of interest should be found in the flow through.

The interaction between a protein and an ion exchanger depends not only on the net charge and the ionic strength, but also on the surface charge distribution of the protein. If there are surface regions with a high concentration of charged groups on the protein, binding can occur even when the overall charge of the protein is zero. This phenomenon is exemplified by yeast phosphoglycerate kinase that binds to a cation exchanger at a pH when its net charge is slightly negative. Moreover, since the chromatographic behavior also depends on protein conformation, some structural changes can affect the separation by IEXC.

The Stationary Phase

The properties of the ion exchanger will also influence the separation. Ion exchangers are usually classified as weak or strong. The name refers to the pKa values of their charged groups (by analogy with weak and strong acids or bases) and does not say anything about the strength of the interaction. Strong ion exchangers have functional groups, e.g. sulphonate and quaternary ammonium, with pKa values outside the pH range in which it is usual to work with proteins (i.e. pH 4-10) and pH changes will therefore not change the charge of the ion exchanger. Contrary, weak ion exchangers have functional ionic groups, e.g. carboxylate and diethyl ammonium, with a limited pH range for their use. Thus, weakly ionizable proteins, requiring a very high pH or very low pH for ionization, can be separated only on a strong exchanger. On the other hand, for more highly charged proteins, weak ion exchangers are advantageous for a number of reasons, including a reduced tendency to sample denaturation, their inability to bind weakly charged impurities and the enhanced elution resolution.

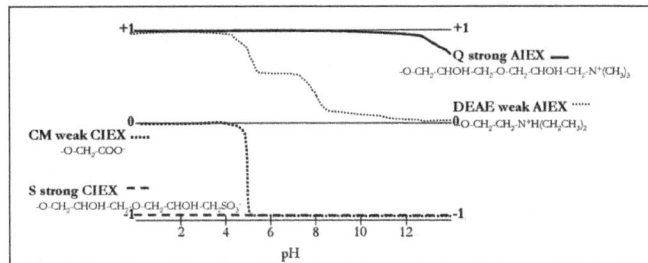

Ionization curves of the most common types of ion exchangers. The strong exchangers Q and S are fully charged at all pH values usable for protein purification.

The Mobile Phase

The pH and the conductivity of the running buffer are of outermost importance since the grade of adsorption is determined by these factors. Low conductivity and a pH that gives the protein an optimal charge have to be chosen.

The choice of buffering ions also influences the separation. For example, the electrostatic attraction between two oppositely charged groups is higher in a hydrophobic environment. In several instances, ammonium acetate buffers have increased the resolution. This buffer has also the advantage of being volatile and can thus easily be removed by lyophilization.

Elution Strategies

Normally, proteins with the same charge as the resin will pass through the column without adsorbing while proteins with the opposite charge will be bound. If the sample components only are differentially retarded, and thereby separated under constant solvent composition, no changes in buffer composition are required. This is termed isocratic elution. In this way only sample volumes much smaller than the total bed volume can be applied and the resolution increases as the square root of the column length.

More often, however, a decrease in affinity is mediated by a change in the buffer, in order to selectively release the proteins from the column. Two general methods are available, changing the pH of the eluting buffer or increasing the ionic strength by addition of NaCl.

If a pH change is used, anion exchangers should be eluted by a decrease in pH to make the adsorbed proteins less negative, whereas cation exchangers are eluted by an increase in pH to make the proteins less positively charged. However, in practice pH-elution is generally not very successful. Since many proteins show minimum solubility in the vicinity of their isoelectric point care and precautions must be taken to avoid precipitation on the column. Moreover, unless having a very high buffering capacity, large pH changes can occur when proteins become eluted. This leads to a less good separation of individual components. However, by employing special systems that maximize buffering capacity, it has been possible to achieve high resolution of proteins by pH elution. This technique is referred to as chromatofocusing.

A more common strategy to achieve elution is to increase the concentration of a non-buffering salt, such as NaCl. These ions compete with the protein for binding sites on the resin. More weakly charged proteins are eluted at lower salt concentrations while the more strongly charged proteins are eluted at higher salt concentrations.

Buffer changes required for elution can be added either in stages, by stepwise elution or continuously, by gradient elution. Stepwise elution is a serial application of several isocratic elution steps, each step consisting of about one bed volume of eluent. Stepwise elution is often used for recovery of a concentrated protein in the breakthrough peak of the displacing buffer. In gradient elution the concentration of the eluting buffer is changed continuously. First the least strongly adsorbed protein is desorbed and at somewhat higher concentration the second protein and so on. Gradient elution generally leads to improved resolution since zone sharpening occurs during elution. The separation obtained is similar to isocratic elution, but because of the continuously increasing elution power, the peaks do not become much broader as the gradient develops. Decreasing the slope of the gradient will lead to a greater separation of the solutes. However, as the slope decreases, the proteins will be more diluted. Due to extended equipment requirements, gradient elution may sometimes not be feasible for large-scale processes. The optimized gradient conditions need, in this case, to be transferred to a series of steps to first elute less retained contaminants, then elute the protein product and finally, release all tightly bound solutes in a wash step.

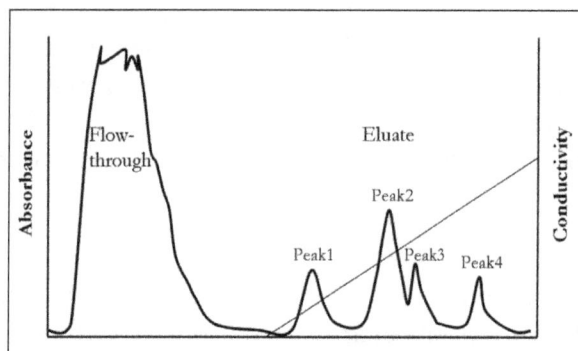

Ion Exchange Chromatogram:
A typical chromatogram from an ion exchange purification using salt gradient elution.

Features

IEXC is one of the more powerful protein purification techniques available and probably the most frequently used chromatographic technique for the separation of proteins, polypeptides, nucleic acids, polynucleotides, and other charged biomolecules. An advantage of this technique is that the elution normally takes place under mild conditions, so that the protein can maintain its native conformation during the chromatographic process.

In general, ion exchangers are more densely substituted than other adsorbents used in protein chromatography and its capacity for protein binding is very high. Its broad specificity also allows for removal of significant impurities such as deamidated forms, endotoxins and unwanted glycoforms. Still, non-specific interactions with proteins due to hydrophobic or other non-ionic interactions are low.

Additional reasons for the success of IEXC are the straightforward separation principle and ease of performance and controllability of the method. Moreover, ion exchanger resins are very robust and can be sanitized in place and used for hundreds of cycles. The main disadvantage of IEXC is its limitations in selectivity.

Chromatographic Methods based on Hydrophobicity

Proteins can be separated by differences in their hydrophobicity using two different methods; Hydrophobic Interaction Chromatography (HIC) and Revered Phase Chromatography (RPC). In both methods the proteins are retained differently by a hydrophobic support depending on their hydrophobicity. The actual nature of the hydrophobic interaction itself is a matter of debate but the conventional wisdom assumes the interaction to be the result of a favorable entropy effect.

Hydrophobicity

The word hydrophob literally means "afraid of water" and refers to the physical property of a molecule that is repelled by water. At present, no universally agreed single measurement exists for hydrophobicity of proteins, although many approaches have been used for the estimation of hydrophobicity of the individual amino acids. Different hydrophobicity scales have been constructed on the basis of free energy transfer for the amino acids from organic solvents to water. Generally, hydrophobic amino acids are those with side chains that lack active groups for formation of hydrogen-bonds with water (e.g. isoleucine, valine, leucine, and phenylalanine) and thus do not like to reside in an aqueous environment. For this reason, one generally finds these amino acids buried within the hydrophobic core of the native protein, or within the lipid portion of the membrane. However, since only a small fraction of the amino acids can be buried, some hydrophobic amino acids will also appear on the surface. The hydrophobicity of native proteins is thus the sum of the hydrophobicities of the exposed side chains and part of the protein backbone.

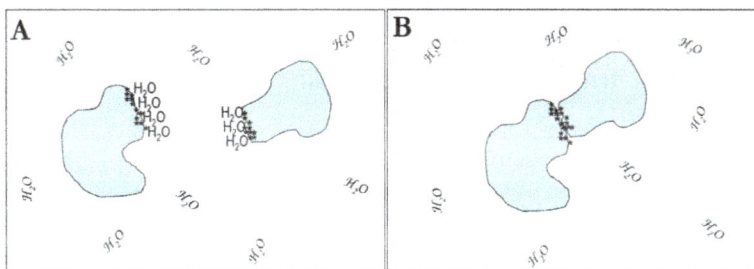

Hydrophobic interaction:
A. Around hydrophobic surfaces the water molecules are highly ordered.
B. When two hydrophobic surfaces interact and shield each other the water molecules are released into the bulk.

In aqueous solution, hydrophobic regions on the protein are covered with an ordered film of water molecules that effectively masks the hydrophobic groups. This increased order of the water molecules leads to a decrease in entropy ($\Delta S < 0$). Thus, hydrophobic compounds will spontaneously associate in an aqueous environment in order to minimize the hydrophobic area that is exposed to the solvent. This is advantageous from an energy point of view, since the ordered water molecules around the non-associated hydrophobic groups will then be released to the more unstructured bulk water. This will lead to an increase in entropy of the system which will result in a net decrease in free energy ($\Delta G = \Delta H - T\Delta S$) of the system. Hydrophobic adsorption of proteins is thus an entropy-driven, thermodynamically favorable process where the driving force is reduction of surface area.

Hydrophobic interactions are now commonly accepted to be of prime importance in many biological systems. It is responsible for the self-association of phospholipids and other lipids to form the biological membrane bilayer and the binding of integral membrane proteins. Moreover, it is a major driving force behind the folding of globular proteins, in the binding of many small molecules to proteins and also in the dynamics of protein motion in solution. Hydrophobic interactions have also been exploited in techniques for separations of proteins.

Hydrophobic Interaction Chromatography

Hydrophobic Interaction Chromatography (HIC) is based on the reversible interaction between a protein surface and a chromatographic sorbents of hydrophobic nature. The proteins are separated according to differences in the amount of exposed hydrophobic amino acids. To facilitate hydrophobic interactions, the protein mixture is loaded on the column in a buffer with a high concentration of salt.

The concept of protein separation under HIC-conditions was outlined by Tiselius already in 1948, when he first reported that proteins are retarded in a buffer containing salts in a socalled salting out chromatography. He noted that "proteins and other substances which are precipitated at high concentrations of neutral salts (salting out), often are adsorbed quite strongly already in salt solutions of lower concentration than is required for their precipitation, and that some adsorbents which in salt-free solutions show no or only slight affinity for proteins, at moderately high salt concentrations become excellent adsorbents". Since then, great improvements have been made in developing the technique. The first matrices of practical use were of a mixed hydrophobic-ionic character. Later, charge-free hydrophobic adsorbents were synthesized and thereby the hydrophobic character of the adsorption could be proved. This led to that Hjertén in 1973 suggested the now generally accepted name of the technique: Hydrophobic Interaction Chromatography. It was also demonstrated that the binding of proteins was enhanced by high concentrations of neutral salts, as previously observed by Tiselius, and that elution of the bound proteins was achieved simply by washing the column with salt-free buffer or by decreasing the polarity of the eluent.

The Stationary Phase

Many types of matrices are suitable for preparing adsorbents for HIC, but the most extensively used has been agarose. When the technique has been adapted to HighPerformance Liquid Chromatography (HPLC), silica and organic polymer resins have also been employed. Since elution is performed at low ionic strengths, the adsorbents should preferably be charge free to avoid ionic interaction between the protein and the matrix.

The most widely used ligands for HIC are linear chain alkanes, with or without a terminal amino group. In general, alkyl (e.g. butyl, octyl) ligands show pure hydrophobic character. Sometimes it can be advantageous to use aryl ligands (e.g. phenyl) which also provide some aromatic (π-π) interactions. A phenyl group has about the same hydrophobicity as a pentyl group, although a phenyl ligand can have a quite different selectivity compared to a pentyl ligand, since aromatic groups on protein surfaces can interact specifically with the aromatic ligands.

The strength of the interaction between a protein and hydrophobic ligands increases with the increase in length of the carbon chain. Ligands containing between 4 and 10 carbon atoms are suitable for most separation problems. However, for proteins with poor solubility in buffers of high salt concentration (e.g., membrane proteins) HIC adsorbents with rather long ligands are recommended.

The protein binding capacities will of course also increase with the amount of immobilized ligands. However, at a sufficiently high degree of ligand substitution the apparent binding capacity of the adsorbent remains constant whereas the strength of the interaction increases. Solutes bound under such circumstances are difficult to elute due to multi-point absorbance. Thus, a rather low ligand density (e.g. in the range of 10-50 μmol/ml gel) is advantageous for preserving protein structure.

The Mobile Phase

HIC requires the presence of certain salt ions, which preferentially take up the ordered water molecules and thereby promote hydrophobic interactions. Both anions and cations can be sorted in a list, called the Hofmeister (lyotropic) series, starting with those that highly favor the interaction to those that will reduce hydrophobic forces.

For anions, the series is: $PO_4^{3-} > SO_4^{2-} > CH_3COO^- > Cl^- > BR^- > NO_3^- > ClO_4^- > I^- > SCN^-$

The cations series is: $NH_4^+ > Rb^+ > K^+ > Na^+ > Li^+ > Mg^{2+} > Ca^{2+} > Ba^{2+}$

The ions at the beginning of this series, called cosmotropes or anti-chaotropes, are considered to exhibit stronger interactions with water molecules and thereby also be water structuring and promote hydrophobic interactions.

The retention mechanism of proteins on HIC matrices has been widely studied but none of the proposed theories has enjoyed general acceptance. The surface increment of the salt has been considered as one of the most important parameters. Depending on its components, different salts affect the surface tension of water differently. Generally, a salt that increases the tension of water will also give rise to an increase in the strength of interaction between proteins and the HIC adsorbent. The strength of molal surface tension follows the series:

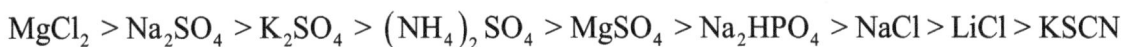

$$MgCl_2 > Na_2SO_4 > K_2SO_4 > (NH_4)_2 SO_4 > MgSO_4 > Na_2HPO_4 > NaCl > LiCl > KSCN$$

However, the effect of salt composition on protein retention is very complex and appears to include other factors such as specific interactions between the protein and the salt, which may change the protein structure and the hydration of the protein. For example, $MgCl_2$ do not enhance the protein binding to hydrophobic stationary phases as much as expected from the surface tension increment.

Ammonium sulphate $((NH_4)_2SO_4)$ and sodium sulphate (Na_2SO_4) are the most utilized salts in HIC. These salts are also known to have a stabilizing influence on protein structure. Unfortunately, ammonium sulphate is instable and forms ammonia gas under basic conditions and therefore the pH should be below 8 when using this salt. Sodium sulphate is suitable as a salting-out agent, but it often causes solubility problems at high concentrations. Since HIC is carried out at high ionic strength the risk for protein precipitation in the system or on the column must be considered. The concentration of salt used for adsorption should therefore always be kept below the concentration that precipitates any protein in the sample. 1 M ammonium sulphate is typically a good starting point for screening experiments. If the substance does not bind then a more hydrophobic medium should be chosen. It is also important that the bound protein can be eluted from the column in a salt-free buffer and with high recovery. If non-polar solvents are required for its elution, a less hydrophobic medium should be employed.

The pH of the buffers used in HIC experiments has a decisive influence on the adsorption of proteins to the adsorbent. Usually, an increase in the pH value (up to 9-10) decreases the hydrophobic interaction between proteins and the hydrophobic ligands, due to the increased hydrophilicity promoted by the change in the charge of the protein. However, some proteins with high pI values bind strongly to HIC matrices at elevated pH values. Since the change in retardation with pH is large for most proteins, it can be worthwhile to test different pH values for adsorption. The only limitation is the stability of the protein to be purified and the stability of the chromatographic matrix (e.g., silica is not stable at high pH).

The temperature dependence of HIC is not simple, although, generally, increasing the temperature enhances protein retention and lowering the temperature generally promotes the protein elution. However, labile proteins should be separated at low temperatures.

Elution Strategies

The favorability of hydrophobic interactions can be weakened if the salt concentration is lowered, and, thereby, the protein can be eluted. Desorption occurs in the order of increasing surface hydrophobicity. Many proteins elute only when the salt concentration is very low. In some cases a decrease of the solvent polarity is also needed. The addition of polarity-reducing compound, such as ethylene glycol, can be made after the salt has been removed from the column, or concomitantly with the decrease of salt concentration. For purification of membrane proteins the addition of detergents (usually 1%), which work as displacers of the proteins, might also be necessary. In some cases the binding is too strong to be useful in a chromatographic process and may be practically irreversible. If organic solvents, detergents or chaotropic agents are required to elute a strongly bound protein, it may lead to protein denaturation.

Simple linear gradients are the first choice for screening experiments, but for optimal separation it might be advantageous to make a gradient shallower in areas where resolution is inadequate. Step-wise elution is often preferred in large scale preparative applications since it is technically more simple and reproducible than gradient elution. It can sometimes also be advantageous in small scale applications since the compound of interest can be eluted in a more concentrated form if the eluting strength of the buffer can be kept high enough without causing co-elution of more strongly bound compounds.

Features

Separations by HIC are often designed using nearly opposite conditions to those used in IEXC. The sample is loaded in a buffer containing a high concentration of salt which makes this method very useful as a subsequent step after proteins are eluted from ionic exchange columns with high salt. The proteins are then eluted from the HIC resin as the concentration of the salt in the buffer is decreased and are then ready for the next purification step without further buffer exchange.

Selectivity is also orthogonal to IEXC as separation is done by a different principle that adds a new dimension to the separation process and does not merely repeat the selectivity of the other. In HIC the proteins are separated primarily by hydrophobic regions on the structure, while differences in charge have little impact on selectivity.

Hydrophobic interaction chromatography is, in general, a very mild method. The structural damage to the biomolecules is minimal, certainly due to the stabilizing influence of salts and also thanks to the rather weak interaction with the matrix. Still, recoveries are often high. Thus, HIC combines the non-denaturing characteristics of salt precipitation and the precision of chromatography to yield excellent activity recoveries.

Reversed Phase Chromatography

For biochemists the name Reversed Phase Chromatography (RPC) can be a bit confusing since it relates to an older technique used in organic chemistry, denoted normal (or polar) phase chromatography, in which the adsorbent is hydrophilic and the liquid in the column is an organic solvent. RPC, as well as the closely related technique HIC, are both based upon interactions between hydrophobic ligands covalently attached to the adsorbent and the hydrophobic patches of molecules that are applied in the aqueous mobile phase. The adsorbents used in the two techniques differ in the way that adsorbents for RPC are an order of magnitude more highly substituted with hydrophobic ligands than in those used for HIC. This leads to that in RPC the hydrophobic interaction is strong enough to adsorb proteins in pure water. However, the very strong interactions that thereby are provided usually require the use of organic solvents and other additives to desorb the protein. This will most often have a denaturing effect on the protein. Still, the basic molecular interactions are very similar to HIC, and RPC may conceptually be regarded as a strong type of HIC or vice versa.

The technique behind RPC was originally developed for the separation of relatively small organic molecules which more or less dissolved in the hydrocarbon phase. It was first in the late 1970s that the use of RPC was applied to purification of polypeptides and it has since then achieved considerable interest due to the high resolving power of the technique. However, the mechanism of the interaction for peptides and proteins deviates distinctly from that of the typical organic molecule. Small molecules are probably subject to partitioning while peptides and proteins, that are rather large in comparison to the traditional organic target molecule, are probably mainly retained by adsorption to the stationary phase, often by multi-point attachment.

The Stationary Phase

The most common base matrix for RPC gels is porous silica beads with modified Si-OH groups to attach the ligand. Silica beads are mechanically strong and also chemically stable in the organic solvents typically used for RPC. However, the coupling between the ligand and the Si-OH group is chemically unstable at high pH values (>7.5). The thereby released Si-OH groups can then be deprotonated which will lead to a mixed chromatography. The aimed hydrophobic interactions between the protein and the ligand will then be combined with ionic interactions between negatively charged silanol groups exposed on the support and the positively charged amino groups on the protein. The effect of this mixed mode retention is increased retention times with significant peak broadening and should thus be avoided.

In order to obtain the strong hydrophobic interaction that is wanted in RPC, the silica particles should be almost completely covered (several hundred µmol/ml gel) with chemically bonded hydrocarbon chains that represent the hydrophobic phase. Any residual silanol groups is believed to contribute to deleterious mixed mode ionic retentions and are therefore reacted with smaller alkylsilane reagents in a process referred to as end-capping.

Small organic molecules behave as if they were dissolved in the hydrocarbon phase and are therefore sensitive to the chain length of the bonded phase. Proteins and peptides, on the other hand, behave as if they were adsorbed to the stationary phase and are much less sensitive in this respect. Shorter carbon chains (C2-C8) are typically lengths used for protein separations, in order to avoid too strong interactions that require higher concentrations of organic modifier for elution. However, longer aliphatic chains (C8-C18) can preferably be used for separation of smaller peptides.

Synthetic organic polymers, e.g. beaded polystyrene, are also available as reversed phase gels. Unlike silica gels, organic polymer packings are stable at pH values up to 12. However, they are usually not as mechanically stable and tend to shrink or swell when exposed to different solvents. These beads have a surface that is itself strongly hydrophobic and, therefore, left underivatised.

The Mobile Phase

The initial mobile phase binding conditions used in RPC are primarily aqueous with a high degree of organized water structure surrounding the column support which is very hydrophobic in nature. This leads to a protein binding that is usually very strong and requires the use of organic solvents in the mobile phase for elution. The organic solvent will lower the polarity and the lower the polarity of the mobile phase, the greater is its eluting strength. Although a large variety of organic solvents can be used, in practice only a few are routinely employed. The two most widely used are acetonitrile and methanol, although acetonitrile is the more popular choice. Isopropanol can be employed because of its strong eluting properties, but is limited by its high viscosity which results in lower column efficiencies and higher backpressures. The relative retention of a particular polypeptide or protein decreases in order of the following series of solvent modifiers at the same volume percentage:

methanol < ethanol < acetonitrile < isopropanol

UV transparency is a crucial property for RPC, since column elution is typically monitored using UV detectors. Peptide bonds only absorb at low wavelengths in the ultra-violet spectrum. Acetonitrile is therefore the almost exclusively choice when separating peptides lacking aromatic amino acids, since it provides much lower background absorbance at low wavelengths.

Addition of an organic solvent modifier to a protein will, in general, have a denaturing effect due to regional disruption of the hydrophobic interactions between nonpolar side chains in the protein and perturbation of the hydrogen bonding characteristics of the protein through disruption of peptide backbone dipoles. Where mobile phase induced denaturation occurs, the three-dimensional structure is disrupted and an increase in surface contact with the stationary phase is anticipated. Consequently, the retention value becomes larger for proteins in the denatured form. Separation in RPC is thus according to differences in the total hydrophobicity, since almost all of the amino acid residues are available for interaction with the stationary phase.

Ion Suppression

The stability of silica-based reversed phase gels dictates that the operating pH of the mobile phase should be below pH 7.5. Thus, a common trick employed routinely with RPC is to prepare the mobile phase with strong acids such as trifluoroacetic acid (TFA) or ortho-phosphoric acid that reduces the pH to between 2 and 3. Low pH conditions will through ion suppression result in the elimination of mixed mode retention effects due to ionizable silanol groups remaining on the silica gel surface. However, there are some further advantages of using a low pH in the mobile phase. Proteins also contain ionizable groups and the degree of ionization will affect their retention in RPC. At low pH conditions, suppression of carboxyl group ionization occurs and the amino groups are essentially fully protonated. Thus, the solute can be considered as a single, averaged ionized species. The isoelectric points of most polypeptides and proteins are above these low pH values. For reasons not yet fully clarified, higher selectivity can be obtained when running below the pI value of a polypeptide or protein in reversed-phase systems. Moreover, low ionic strengths can be used at a low pH value, which facilitates recovery and usually gives a better peak shape and more reproducible retention.

Ion Pairing

Trifluoroacetic acid (TFA) in the mobile phase can also be attributed the effect of an ion pairing agent. Ion pairing agents are thought to bind to the sample molecules by ionic interactions, which results in the modification of the net hydrophobicities. Thereby the retention times of proteins can be modified and the selectivity changed. In some cases the addition of ion pairing agents to the mobile phase is an absolute requirement for binding of the solute to the reversed phase gel. However, their greatest advantage is in affecting selectivity and thereby improving the chances for complete resolution of sample components. The retention behavior of the sample components may be affected by both the type and concentration of ion. Many other alkanoic acids have been proven effective mobile-phase additives, although TFA is the most commonly used. It is preferable to use a volatile acid that can be removed readily by lyophilization.

Elution Strategies

While proteins strongly adsorb to the surface of a reversed phase matrix under aqueous conditions, they desorb from the matrix within a very narrow window of organic modifier concentration. A typical biological sample usually contains a broad mixture of biomolecules with a correspondingly diverse range of adsorption affinities. The only practical method for reversed phase separation of complex biological samples, therefore, is gradient elution. This is usually done by decreasing the polarity of the mobile phase by increasing the percentage of organic modifier in the mobile phase.

Features and Applications

The strong adsorption and the organic modifiers needed for desorption in RPC usually leads to protein denaturation. For successful preparative purification, either an inactive protein must be adequate for the purpose or, alternatively, the protein must be sturdy enough to withstand the rigors of the environment during chromatography, or, if not, a return to the correct tertiary structure must be easily attained. For proteins, this technique is therefore mostly used for purity check and quality control analyses, when recovery of activity and tertiary structure are not essential. For smaller proteins (Mw <30,000), the denaturation effects are often minimal or rapidly reversible, and RPC can therefore be successfully used to isolate them in a biologically active form. Shorter polypeptides usually have no real secondary structure to preserve, and so cannot denature in the conventional sense, why RPC is often the purification of choice in these cases.

Reversed Phase High-Performance Liquid Chromatography

With the advent of media and instrumentation that can withstand high pressure, RPC has been most typically used as a High-Performance Liquid Chromatography (HPLC) technique, where its inherent robustness is especially advantageous. Due the high resolving power of the technique RP-HPLC has found wide application for the separation, purification, and analysis of small polypeptides and proteolytic fragments. For analytical separations of complex protein fragmentation patterns RP-HPLC is the only method of choice, since it is the only technique possessing sufficient resolving power to resolve the digested protein sample completely. RP-HPLC is also an extremely useful tool for final polishing and for isolation of proteins and peptides prior to mass spectrometry analysis as well as microsequencing using the Edman degradation procedure.

Affinity Chromatography

The biological function of proteins often involves specific interactions with other molecules, called ligands. These interactions might occur with low molecular weight substances such as substrates or inhibitors but do particularly occur with other proteins. An interacting protein has binding sites with complementary surfaces to its ligand. The binding can involve a combination of electrostatic or hydrophobic interactions as well as short-range molecular interactions such as van der Waals forces and hydrogen bonds. Affinity chromatography owes its name to the exploitation of these various biological affinities for laboratory purification of proteins. A specific ligand is then covalently attached to an inert chromatographic matrix. The sample is applied under conditions that favor specific and reversible binding of the target protein to the ligand. Since only the intended protein is adsorbed from the extract passing through the column, other substances will be washed away. To elute the target molecule the experimental conditions are changed so that the protein-ligand interaction is broken.

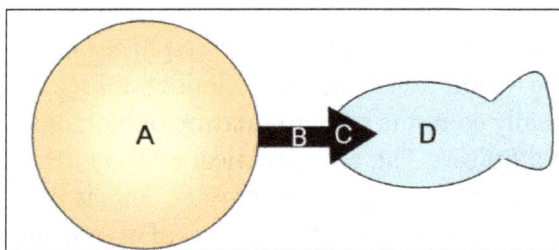

An affinity matrix binding to its target protein:
A. The bead. B. The Spacer arm. C. The ligand. D. The target protein.

The technique was originally developed by Cuatrecasas, Wilchek and Anfinsen in 1968 for the purification of enzymes but it has since been extended to receptor proteins, immunoglobulins, glycoconjugates, nucleotides and even to whole cells and cell fragments. Applications of the technique are limited only by the availability of immobilized ligands.

The term affinity chromatography referred originally to the use of an immobilized natural ligand, which specifically interacts with the desired protein, but has then been given quite different connotations by different authors. Sometimes it is very broad and includes all kinds of adsorption chromatography techniques based on non-traditional ligands, and is thus used in a more general sense of attraction. In other cases it refers only to specific interactions between biologically functional pairs which interact at natural binding sites.

The Stationary Phase

For affinity chromatography applications, the ideal gel material should have certain characteristics. First of all, it must possess suitable chemical groups to which the ligand can be covalently coupled and have a relatively large surface area available for attachment. The harsh conditions that are used during derivatization demand that the matrix must be both chemically and mechanically stable. It must also be inert in the solvents and buffers that are employed in the process, which can be rather harsh, especially during elution of the protein. Hydrophilic and neutral matrices are preferred, to prevent the proteins from interacting nonspecifically with the gel matrix itself. The matrix should be macroporous to accommodate free interaction of large proteins with the ligands but it must also exhibit good flow properties.

Ever since its introduction, agarose has been the most popular base for affinity matrices. A contributing reason for this popularity is that there were early simple and convenient coupling methods developed for agarose, and even commercially available preactivated matrices.

A number of synthetic organic and inorganic porous bead matrices are also available, e.g. cross-linked dextrans, polystyrene, polyacrylamide, cellulose, porous glass and silica.

Ligands

Successful affinity purification requires a biospecific ligand that can be covalently attached to the chromatographic matrix. Ligands can be extremely selective and bind to only a single or a very small number of proteins. Examples are antibodies, protein receptors, steroid hormones, vitamins, and certain enzyme inhibitors. Some ligands are less selective and bind to a group of closely related compounds with similar chemical characteristics. However, these interactions have also proved to be extremely helpful in solving many separation problems. Good examples are staphylococcal protein A and G ligands that are group selective for immunoglobulins.

The coupled ligand must be able to form reversible complexes with the protein to be isolated. The stability of the complex should be high enough for the formation of complexes at least sufficient for retardation in the chromatographic procedure. However, after washing away unbound material it is important that it is easy to dissociate the complex by a simple change in the medium, without irreversibly affecting the protein to be isolated or the ligand.

For the preparation of the affinity adsorbent the ligand should be compatible with the solvents used during the coupling procedure. In this sense the best type of ligand for affinity chromatography would be a synthetic molecule that is stable, safe and inexpensive. Some dyes have been used in this way, but the problem so far has been a lack of specificity. Protein ligands usually provide higher selectivity but are not ideal for production, since they can be irreversibly denatured by cleaning solutions. Moreover, proteins are expensive and must be pharmaceutically pure before being bound to the column.

It is essential that the ligand possesses at least one functional chemical group by which it can be immobilized to the matrix. The most common of such groups are amines, thiols, carbohydroxides and hydroxyl groups. It is also important that the functional group that is used for coupling is non-essential for its binding properties to assure that the ligand retains its specific binding affinity for the target molecules.

Spacer Arms

To prevent that the attachment of the ligand to the matrix interferes with its ability to bind the target molecule, it is generally advantageous to interpose a spacer arm between the ligand and the matrix. In this way the ligand is distanced from the surface of the matrix which reduces steric hindrance of the binding that can occur when the ligand is bound directly to the bead. Spacers are most important for small immobilized ligands and are generally not necessary for macromolecular ligands. The optimum length of a spacer arm is 6-10 carbon atoms or their equivalent. Spacer arms should neither chemically or structurally affect the sample or the ligand.

Ligand Coupling

In general, the procedure for immobilization of a ligand consists of three steps. First the matrix is activated to make it reactive toward the functional group of the ligand. Thereafter the ligand is co-valently coupled through some chemical reaction. Finally, residual unreacted groups are blocked by a large excess of a suitable low molecular weight substance such as ethanolamine. This provides a higher degree of certainty that all binding will be between the ligand and the sample.

Methods for activation of an affinity adsorbent are varied and dependent on the chemistry of the ligand and the adsorbent itself, and also whether a spacer arm is required. Some commonly used immobilization procedures are shown in table. The matrix must be activated in such a way that quite gentle chemistry allows covalent attachment of a ligand. The activation normally consists of the introduction of an electrophilic group into the matrix. During ligand coupling this group reacts with nucleophilic groups, such as amino, thiol and hydroxyl groups on the ligand. Alternatively, a matrix activated with nucleophilic groups, e.g. thiol, can be used to immobilize a ligand containing an electrophilic group, although such an approach is less common.

Reagent for activation	Activated matrix	Functional group on ligand	Coupled ligand
CNBr	—O— C≡N	H₂N-ligand	—O—C—NH-Ligand (NH₂⁺)
Epichloro hydrine	—O—CH₂—CH—CH₂ (O epoxide)	HS-ligand	—O—CH₂—CH—CH₂—S-Ligand (OH)
Epichloro Hydrine + 6 aminohexanoic acid + N-hydroxy succinimide (NHS)	—O–CH₂-CH–CH₂-NH—(CH₂)₅CO–O–N (OH)	H₂N-ligand	—O–CH₂-CH–CH₂-NH—(CH₂)₅—C-NH-Ligand (OH)
N,N' carbonyl diimidazole (CDI)	—O—C—N (O)	H₂N-ligand	—O—C—NH-Ligand (O)
Tosyl chloride	—O—SO₂——CH₃	H₂N-ligand	—NH-Ligand

Overview of some commonly used immobilization procedures.

The most common method of attachment involves treatment with cyanogen bromide (CNBr) which reacts with hydroxyl groups in polysaccharide matrices to produce a reactive support. CN-Br-activated matrices are suitable for ligands such as polypeptides and proteins, since they react swiftly in weakly alkaline conditions (pH 9-10) with primary amines to give principally an isourea derivative. Unfortunately, the isourea bond is positively charged at physiological pH (pKa ~ 9.5) and is thus imparting anion exchange properties to the adsorbent. On the other hand, many of the ligands that are attached are negatively charged, so the isourea derivative may cancel possible cation exchange effects. However, the CNBr technique is not ideal for single point attached ligands since the isourea bond can be cleaved rather easily. Since CNBr is extremely toxic and releases HCN on acidification, commercially activated matrices are recommended.

The matrix can also be activated using other chemical reactions. For example, polysaccharide matrices can be activated by epichlorohydrins or bisepoxiranes which introduce an epoxy group as the active electrophile. Except reaction with primary amines, epoxy groups react rapidly with sulf-hydryls and also slowly with hydroxyls. This is especially useful since is provides the possibility of coupling sugar ligands. An interesting characteristic when using bisoxirane is that the ligands will be provided automatically with a 12 atom long hydrophilic spacer arm which may be desirable in certain applications.

After attachment of for example an alkanoic acid as a spacer arm to the matrix, the terminal carboxyl group of this moiety can be further activated to form active N-hydroxysuccinimide (NHS) esters. NHS-activated matrices are now commonly used, since ligands containing primary amino groups can react directly with this active ester to form a chemically very stable amide linkage.

Polysaccharide matrices can also be activated using less noxious carbonylating agents such as N,N' carbonyldiimidazole (CDI) to give reactive imidazole carbonate derivatives. These in turn will at alkaline pH readily react with ligands containing primary amines under the formation of carbamates. The carbamate bond is very stable and non-charged under conditions usually employed for affinity chromatography. However, this linkage gives minimal spacing from the matrix.

A method that does not introduce any spacer at all uses organic sulfonyl halides e.g. tosyl chloride or the more reactive tresyl chloride. These agents react with hydroxyl groups to form sulfonates, which are themselves good leaving groups, that allow binding of nucleophiles on the ligands directly to the hydroxyl carbon.

Other adsorbents containing hydroxyl groups, e.g. some cellulose and silica matrices, can also be activated successfully using organic sulfonyl halides. Silica gels require generally more considerations before activation, since the negatively charged silanol groups can be changed by chemical modification. Matrices having amide groups, such as polyacrylamide, can also be activated using glutaraldehyde and hydrazine methods that do not work directly on polysaccharide matrices.

Many pre-activated matrices, prepared using these coupling reagents to facilitate the coupling of specific types of ligand, are available commercially. Several supports of the agarose, dextran and polyacrylamide type are also commercially available with a variety of spacer arms and ligands pre-attached ready for immediate use.

The Mobile Phase

In affinity chromatography, a certain degree of care must be used to make binding between target molecule and ligand as opportune as possible. The ideal binding buffer conditions are optimized to ensure that the target molecules interact effectively with the ligand and are retained by the affinity medium while nonspecific interactions are minimized. In most cases the binding buffer is also used to wash unbound substances from the column without eluting the target molecules. Variations in flow rate can exhibit monumental effects on the success of affinity chromatography. If the sample is pumped too quickly proper binding may not take place.

In the elution phase buffer conditions are changed to break the interaction between the target molecules and the ligand and thereby eluting target molecules from the column. A property that needs special consideration is the association strength between ligand and counterligand. If it is too weak there will be no adsorption, whereas if it is too strong it will be difficult to elute the adsorbed protein. It is always important to find conditions that promote the dissociation of the complex without at the same time destroy the active protein and degrade the matrix. This is often the major difficulty with affinity chromatography.

Elution Strategies

The principle of desorption is to change the binding equilibrium for the adsorbed substance from the stationary to the mobile phase. This is done by altering the conditions of the solution that the ligand and biomolecule are binding in, so that binding is no longer favorable. This can be achieved in many different ways, either specifically or non-specifically. It is important that the elution buffer work quickly and do not change the function or activity of the desired protein.

Ligand-protein interaction is often based on a combination of electrostatic and hydrophobic interactions and hydrogen bonds. Agents that weaken such interactions might be expected to function as effective non-specific eluants. Careful consideration of the relative importance of these three types of interactions for the stability of the bound protein will help in the choice of a suitable eluant.

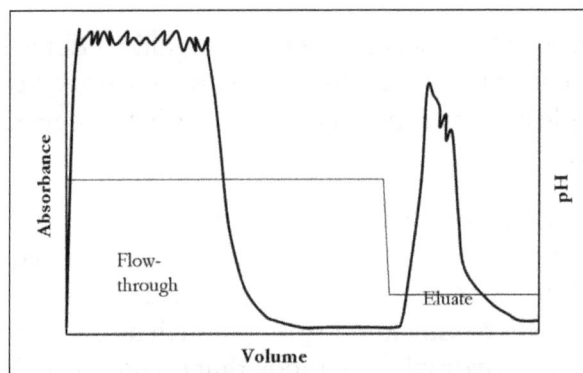

Affinity chromatogram:
A typical chromatogram from an affinity purification where elution is accomplished with a low pH buffer.

A pH change results in a change in the state of ionization of groups in the ligand and the target molecule and might be critical for binding. Therefore, the most frequently used method for eluting strongly bound substances non-specifically is by decreasing the pH of the buffer. Often a pH value around 2 is needed for desorption, but the chemical stability of the matrix, the ligand and the adsorbed substance determine how low pH values it is possible to use. Sometimes an increase in pH using for example ammonium hydroxide can also be effective. If elution is achieved by a pH change, it might be necessary to neutralize the pH of the collected fractions directly after elution to minimize the risk for protein denaturation. An increase in the ionic strength of the buffer can be useful for elution of proteins bound by predominantly electrostatic interactions. Such interactions typically dominate binding to dye columns. 1M NaCl is frequently used for this purpose.

When the binding is dominated by strong hydrophobic interactions, rather drastic methods of elution sometimes have to be used, such as reducing the polarity or including a chaotropic salt or denaturing agent in the buffer. This type of elution is typical for immunosorbent based on immobilized polyclonal antibodies.

In specific elution, bound proteins are desorbed from the ligand by the competitive binding of the eluting agent that binds either to the ligand or to the target molecule.

The addition of a free ligand works by challenging the matrix-bound ligand for the target molecule. If the target molecule forms a stronger binding with the free ligand, it will desorb from the matrix-bound ligand and elute with the free ligand.

The addition of a competitor works almost exactly like the free ligand, but instead of binding to the target molecule, the competitor binds to the matrix-bound ligand. By having a higher affinity for the matrix-bound ligand, or being in a higher concentration, the competitor replaces the target molecule which is then eluted.

Features

Affinity chromatography is a purification technique that simplifies the isolation process by using pre-existing ligand-binding relationships that are developed to be highly specific. As a consequence, affinity chromatography is theoretically capable of giving very high purification power, even from complex mixtures, in one simple process step. Even when the protein of interest is a minor component of a complex mixture affinity chromatography can be successful since the concentrating effect enables large volumes to be processed.

Another feature of affinity chromatography is that active molecules often can be separated from denatured or functionally different forms, since the technique relies on functional properties.

Applications

Affinity chromatography has a broad range of applications for protein purification. Some frequently used biological interactions are further described.

Immunoaffinity

The high specificity of antibodies makes them extremely useful as ligands for affinity chromatography, especially when the target molecule has no immediately apparent complementary binding substance other than its antibody. In immunoaffinity, the immobilized protein is known as an immunoadsorbent. Traditional immunoadsorbents based on polyclonal antibody preparations have largely been replaced by adsorbents based on monoclonal antibodies. Modern hybridoma technology and suitable screening methods makes it possible to obtain antibodies of virtually any desired specificity and affinity which can be immobilized for use in purifying the antigen. However, one serious disadvantage with monoclonal antibodies is their high cost. Sometimes affinity chromatography is used in the opposite direction and antigens are used as affinity ligands to allow purification of a specific antibody.

The very tight binding between the antibody and its antigen $\left(Kd = 10^{-8} \text{ to } 10^{-12}M\right)$ often requires harsh conditions for elution of the bound protein. This may inactivate the target protein or even destroy part of the immunoaffinity ligand. The use of chaotropic agents or lowering the pH to about 3 may be useful to avoid denaturation. Moreover, other proteins in the sample preparation, e.g. proteases, may inactivate antibody ligands or bind nonspecifically to them. Thus, crude extracts should never be applied directly to columns packed with expensive adsorbents based on monoclonal antibodies. Moreover, some of the antibodies often leaches from the column during purification and must then be removed from the eluted target protein.

Purification of Immunoglobulins

One form of affinity chromatography commonly used in production uses Protein A or Protein G as ligands for purification of immunoglobulins from various species. These ligands are cell wall proteins from Staphlococcus or Streptococcus with high affinity for the constant region of IgG. Binding is usually achieved at physiological pH values and lower pH values of around 3 is usually needed for elution, which can denature or partially denature the IgG. However, this can be partly avoided by selecting clones of monoclonal IgG's that are stable, and immediately raising the pH while the IgG is being collected. The Protein A and G ligands are both rather stable.

Purification of Glycoconjugates

Lectins are a group of proteins that have the ability to interact specifically and reversibly with certain carbohydrates or a group of carbohydrates. Immobilized lectins are invaluable tools for separation of glycoconjugates such as glycoproteins, polysaccharides, glycolipids and even whole cells containing glycoproteins with specific carbohydrate structures on their surface membrane. Actually, lectins were originally used for agglutination of erythrocytes in the process of blood typing. Nowadays, lectins are valuable in proteomic studies of glycopeptides since lectins are able to recognize them even after proteolysis.

The most widely used lectins are those from leguminous plants, owing to their abundance. Each lectin has its own preferences for binding to a certain sugar sequence. By far the most widely used lectin is Concanavalin A, which interacts strongly with mannose and glucose residues. Most glycoproteins have exposed mannose residues, since these moieties often are covalently attached by posttranslational modification during excretion from the eukaryotic cell, and can thus be purified using a Concanavalin A ligand. There are also other lectins commercially available, e.g. those with specificities towards galactose and N-acetyl galactosamine.

Substances bound to the immobilized lectin can be displaced with a competitive substance such as the monosaccharide for which the lectin has an affinity. The binding can also be reversed with a borate buffer, which forms a complex with glycoproteins. Alternatively, nonspecific elution can be achieved by a change in pH or by the addition of a reagent such as ethylene glycol to reduce hydrophobic interactions to the ligand.

Purification of DNA-Binding Proteins

DNA binding proteins form an extremely diverse class of proteins including those responsible for the replication and orientation of the DNA such as histones, nucleosomes and replicases and those involved in transcription such as RNA/DNA polymerases, transcriptional activators and repressors and restriction enzymes. However these proteins all share a single characteristic, their ability to bind to DNA.

Among other things, this property enables group specific affinity purification using heparin as a ligand. Heparin is a highly sulphated glycosaminoglycan that has two modes of interaction with proteins. In one mode it mimics the polyanionic structure of the nucleic acid and thereby interacts strongly with DNA binding proteins. However, heparin can also acts as an affinity ligand and bind a very wide range of other biomolecules such as the coagulation factor antithrombin.

Of course DNA binding proteins can also be isolated using the specific DNA sequence to which it binds as a ligand. For example, different transcription factors, that are extremely important to understand gene expression, preferably are analyzed in this way. It is important that DNA columns are handled and cleaned with care, to minimize disruption by DNases.

Purification of Receptor Proteins

Receptor proteins can also be purified by affinity chromatography by use of the effector in question as the immobilized ligand. Examples of receptor proteins that are routinely purified in this way are steroid, esterogen, transferrin and androgen receptors.

Purification of Enzymes

For isolation of an enzyme, the ligand may be the substrate, a competitive reversible inhibitor or an allosteric activator. For example, adenosine monophosphate (AMP) can be immobilized and used to bind proteins exhibiting an affinity for AMP, ADP, or ATP. Another example is the synthetic inhibitor para-aminobenzamidine which is frequently used as the affinity ligand for removal of trypsin-like serine proteases and zymogens from cell culture supernatants, bacterial lysates or serum.

Purification by the use of Synthetic Dye Ligands

A number of synthetic triazine dyes have been explored for protein applications. The blue textile dye Cibacron Blue F3G-A is an analog of adenylyl-containing cofactors. Consequently, when this biomimetic is coupled to an adsorbent it can be used to purify a very wide range of enzymes requiring such cofactors, including both NAD+ and NADP+ dependent enzymes. However, the matrix will also acts as an aromatic anionic ligand and bind strongly to a wide range of other proteins including albumin and some lipoproteins, although by electrostatic and/or hydrophobic interactions in a less specific manner. This lack of specificity and questions concerning purity, leakage and toxicity has contributed to a limited use of these matrices.

Isolation of Cells

Another valuable development of affinity chromatography is its use for the separation of a mixture of cells into homogeneous populations. The technique can rely on different properties of the cell surface, for example the chemical nature of exposed carbohydrate residues or specific membrane receptor-ligand interactions.

Isolation of Nucleotides

Affinity chromatographic techniques can also be used for purification of oligonucleotides. Immobilized single-stranded DNA can preferably be used to isolate complementary RNA and DNA. Moreover, messenger RNA is now routinely isolated by selective hybridization on poly(U)-ligands by exploiting its poly(A) tail.

Affinity Tags

Although there are several affinity systems available, most target proteins lack a suitable affinity ligand usable for capture on a solid matrix. A way to circumvent this obstacle is to genetically fuse the gene encoding the target protein with a gene encoding a protein with suitable affinity, denoted a purification tag. When the chimeric protein is expressed the tag allows for specific capture of the fusion protein. Fusion protein purification systems can be based on several kinds of interactions such as protein-protein interactions, enzyme-substrate interactions, protein-carbohydrate interactions and protein-metal interactions. Examples of commonly used purification tags are; Glutathione S-transferase (GST), the maltose-binding protein (MBP), the FLAG tag, the S-tag, the Z-domain, the calmodulin binding peptide, the BioTag, the Strep tag and the His-tag.

Immobilized Metal Affinity Chromatography

Immobilized metal affinity chromatography (IMAC) relies on the formation of weak coordinate bonds between immobilized metal ions and some amino acids on proteins, mainly histidine residues. Since it does not operate through a biospecific interaction in its general sense it is usually called a pseudo-affinity technique.

The foundation of IMAC was introduced by Porath already in 1975 but it has lately found deeper interest thanks to development of more stable adsorbents and the evolving genetic engineering methods.

Metal Ion Affinities

Although still not fully understood, the interaction used in IMAC depends on the formation of coordinated complexes between metal ions and electron donor groups on the protein surface. Some amino acids are especially suitable for binding and histidine is the one that exhibits the strongest interaction, as electron donor groups on the imidazole ring in histidine readily form coordination bonds with the immobilized transition metal. Cysteines can also contribute to binding if free sulfhydryl group are available in the appropriate, reduced state. Although also aromatic side chains of Trp, Phe and Tyr can interact with metal ions the actual protein retention in IMAC is based primarily on the availability of histidyl residues. Since many proteins contain these amino acids, it might be expected that all proteins are capable of binding to metal chelate columns. However, the residues must be located at the surface of the protein for successful coordination and the strength of interaction will depend on the number of such coordinations.

The Stationary Phase

Basically, the requirements of the support in IMAC are the same as also apply to affinity chromatography; easy to derivatize, exhibit no unspecific adsorption and have good physical, mechanical and chemical stability. Beaded agarose is the support predominantly used.

A suitable spacer arm plus a simple chelator is then attached to the matrix. Some commonly used chelators are imino diacetate (IDA), nitrotriacetic acid (NTA) and tris(carboxymethyl) ethylene diamine (TED). Electron-donor atoms (N, S, O) present in the chelators are capable of coordinating metal ions and forming metal chelates. The chelator can be bidentate, tridentate, etc, depending on how many coordination bonds of the metal ion it will occupy. It is important that the metal ion, once bound, does not have all coordination sites occupied by the chelator. The spare coordination sites are initially weakly occupied by water or buffer molecules, but these can easily be exchanged by more strongly complexing electron-donor groups on the target protein. In general, tetradentate ligands will provide more stable metal chelates than the tridentate ligand but the formed metal chelator will then exhibit lower protein binding due to the loss of one coordination site.

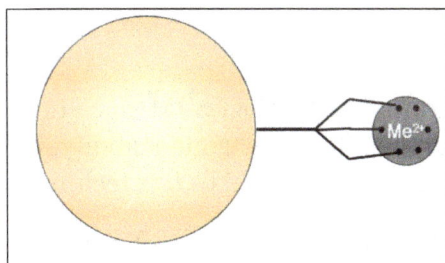

An IMAC matrix:
A bead with an attached chelator that coordinates a metal ion.

The apparent affinity of a protein for a metal chelate depends of course also strongly on the metal ion involved in coordination. Most commonly used are divalent ions of the transition metals Fe^{2+}, Co^{2+}, Ni^{2+}, Cu^{2+}, and Zn^{2+}, but also trivalent metal ions such as Fe^{3+} and Al^{3+} are sometimes used. In the case of the IDA chelator, the affinities of many retained proteins and their respective retention times are in the following order: $Cu^{2+} > Ni^{2+} > Zn^{2+} > Co^{2+}$.

Chelating ligands are relatively inexpensive and capable of high metal ion loadings, which permit high protein-binding capacities. Moreover, the matrices can be re-used many times and can easily be regenerated by adding a buffer containing the specific metal ion. The ligands are stable over a wide range of temperatures and solvent conditions although reducing and chelating agents must be avoided as these readily displace chelated metal ion. The loss of metal ions is more pronounced at lower pH values. Apart from leading to reduced adsorption capacity, metal ions that leak from the sorbent can cause damage to the target proteins by metal-catalyzed reactions.

The Mobile Phase

Metal chelate columns are likely to possess an overall charge since the chelate is negative, the metals are positive, and they do not always cancel out. Therefore buffers of relatively high ionic strength (containing 0.1 to 1.0 M NaCl) are preferably used to reduce non-specific ionic adsorption.

Since adsorption of a protein to the IMAC support is only possible when the imidazole nitrogens in the histidyl residues are not protonated the binding buffer must have a neutral or slightly alkaline pH.

Elution Strategies

Elution of the target protein from an IMAC resin can be achieved by protonation of the histidine residues by decreasing the pH to 4-5. For proteins sensitive to low pH, competitive elution with imidazole at nearly neutral pH is more favorable. However, the addition of imidazole may cause precipitation of the sample. Application of a strong chelating agent, such as EDTA, also results in elution of the bound proteins. However, this treatment will also destroy the binding properties and the column must be recharged with metal ions prior to the next separation.

Unfortunately, there are a couple of endogenous E. coli proteins that have metal binding capacity, which impair the specificity of IMAC. However, a high salt buffer can reduce nonspecific binding and a pre-elution step with a low concentration of imidazole or EDTA can be used to elute contaminating proteins containing histidine residues.

Features

Compared to other affinity separation technologies IMAC can not be classified as highly specific, but only moderately so. However, benefits like ligand stability, high protein loading, mild elution conditions, simple regeneration and low cost makes this technique highly useful anyway.

Another distinct advantage of IMAC over biospecific affinity techniques is its structure independent interaction that makes it applicable under denaturing conditions. This is often necessary when recombinant proteins are highly expressed in E. coli in the form of inclusion bodies.

However, for the production of therapeutic proteins in substantial quantities, IMAC is not the preferred technique due to problems with reproducibility, leaching of metal ions and contamination of host cell proteins, endotoxins, DNA and viruses.

Applications

Numerous natural proteins contain histidine residues in their amino acid sequence. However, histidines are mildly hydrophobic and only few of them are located on the protein surface. Initially, IMAC techniques were used for isolation of proteins with naturally present, surfaceexposed histidine residues. Since then, genetic engineering of histidine affinity handles attached to the N- or C-terminus of recombinant proteins has revolutionized the IMAC technique. Numerous histidine tags, from very short ones, e.g. HisTrp, to rather long extensions, containing up to ten histidine residues, has been employed. Nowadays, the His6 tag, consisting of six consecutive histidine residues, is one of the most commonly used tags for facilitated purification of recombinant proteins. The number of free coordination sites of the immobilized metal ion determines how many histidines that are able to bind concurrently. However, more histidine residues will enhance the probability for the correct orientation desired for coordination interactions. Since numerous neighboring histidine residues are uncommon among naturally occurring proteins, oligo-histidine affinity handles form the basis for high selectivity and efficiency, often providing a one-step isolation of proteins at over 90% purity. An ideal affinity tag should enable effective but not too strong binding, and allow elution of the desired protein under mild, non destructive conditions. In the case of recombinant E. coli many host proteins strongly adhere to the IMAC matrices and are eluted with the target proteins. Therefore, new approaches for selecting improved histidine tags have focused on elution of the target protein in the contaminant-free window.

Size Exclusion Chromatography

A slightly different process setup is used in Size Exclusion Chromatography (SEC). All previously discussed techniques separates proteins according to some property that provides interaction to certain matrices and thereby retardation from proteins without this property. In SEC, the matrix consists of porous particles and separation is instead achieved according to size and shape of the molecules.

The technique is sometimes also referred to as gel filtration, molecular sieve chromatography or gel-permeation chromatography.

Molecular Sieving

For separation in SEC, molecular sieve properties of a variety of porous materials are utilized. SEC matrices consist of a range of beads with slightly different pore sizes. The separation process depends on the different ability of various proteins to enter all, some or none of the channels in the porous beads. Molecules running through a SEC column have to solve a maze which becomes more complex the smaller the molecule is, as the small molecules have more potential channels that they can access. Larger molecules on the other hand, are for steric reasons excluded from the channels, and pass quickly between the beads. The detour through the channels will thus retard smaller molecules in comparison to larger proteins.

Although the separation in SEC is generally assumed to be according to molecular weight, it is more accurate to claim that it is achieved by the differential exclusion or inclusion within porous particles. The ease of diffusion is dependent on the hydrodynamic volume, which is the volume created by the movement of the molecule in water. The difference between hydrodynamic volume and molecular weight is shape. Proteins tend to be globular molecules while DNA or polysaccharides tend to be linear molecules. Linear molecules have much larger hydrodynamic volumes than globular molecules, so a 10,000 MW DNA molecule will elute much earlier than a 10,000 MW protein.

The Stationary Phase

In SEC the selectivity is aimed to be dependent solely of the inherent porosity of the material. Therefore the support chemistry used for SEC matrices is chosen to reduce adsorptive properties.

The matrices used in SEC are often composed of natural polymers such as agarose or dextran but may also be composed of synthetic polymers such as polyacrylamide. Gels may be formed from these polymers by cross-linking to form a three-dimensional network. Different pore sizes can be obtained by slightly differing amounts of cross-linking. The degree of crosslinking will define the pore size. The first commercial SEC media, Sephadex, composed of dextran that was crosslinked with epichlorohydrin. Many gels are now commercially available in a broad range of porosities. The surface of theses supports contains predominantly hydroxyl groups and provides a good environment for hydrophilic proteins. However, the hydrophilicity is somewhat reduced by the introduction of crosslinking reagents. Some polymers, like agarose, are able to spontaneously form gels under the appropriate conditions.

Macroporous silica has also been employed in SEC but must then be coated with a hydrophilic layer to prevent denaturation of proteins.

All molecules larger than the pore size are completely excluded from the pore channels and thereby unretained and elute together. Hence, the size of the pore in the gel can be adjusted to exclude all molecules above a certain size. The size of the pores in the matrix will also determine the rate of molecules that can enter the pores. The average residence time in the channels depends on size and shape of the molecule and different molecules will therefore have different total transit times through the column. Resins are usually classified based on their capacity to separate different sizes of a hypothetical, globular protein. The upper range is the range at which larger molecules are completely excluded from the channels and thus allowing for no separation. The lower range is the range at which small molecules are able to enter all channels and molecules that are even smaller will not have any extra channels to access and there is thus no selection below this size. The linear range between these two extremes is what usually is reported for each matrix.

It is important to select a gel with suitable separation range. Smaller pores are generally used for rapid desalting procedures or for peptide purification. Larger ones are used for small proteins, while very large ones are used for biological complexes. For optimal and reproducible separation it is important that the pores have a carefully controlled range of sizes. However, due to limitations in the synthesis of porous polymers, uniformity of pore distributions within the gel is problematic, especially for those with very large pores.

The Mobile Phase

In contrast to other types of media the selectivity of a SEC matrix is not adjustable by changing the composition of the mobile phase. Optimally there is no adsorption involved, and the mobile phase should be considered as a carrier phase and not one which has a large effect on the chromatography. However, the sample may require a buffer solution with a well defined pH and ionic composition chosen to preserve the structure and biological activity of the substances of interest.

Elution Strategies

Since molecules are not adsorbed but only retarded on a SEC column the proteins are eluted isocratically and will elute in order with the largest first. A single buffer is used, which means that no gradient pumping systems are needed.

Features

The SEC separation method gives the least resolution with the lowest capacity and largest dilution of the sample with respect to all other forms of chromatography. Still, the method is frequently used due to its simple performance and some features not found in other techniques. The principal advantage of SEC is its gentle non-interaction with the sample, enabling high preservation of biological activity. Moreover, since the separation is not dependent on any adsorptive property of the molecule, SEC provides a method for separating multimers that are not easily distinguished by other chromatographic methods.

Since separation starts directly as the sample is applied on the column, the cross section must be large enough to cope with the desired sample volume. However, the length of the column is also significant since it affects both resolution and process time. Consequently, SEC is preferably used as a final polishing step when sample volumes have been reduced.

Applications

Size exclusion has a wide range of applicability both in preparative and analytical protein purification. Preparative SEC may be arbitrarily divided into buffer exchange and fractionation. The porosity of the matrix is chosen according to the aimed purpose.

Buffer Exchange

Buffer exchange, or desalting, refers to the situation where low molecular weight components of the sample, e.g. salt molecules, are exchanged for another buffering substance or solvent. SEC can preferably be used for buffer exchange, as the low molecular weight salt molecules are easily separated from the much larger proteins. The porosity of the gel is in this case selected to exclude the proteins to be desalted. Since the protein is eluted already in the void volume, while the contaminating solutes of lower molecular mass are retained in the pore channels, desalting may be carried out at high flow rates without impaired resolution. Moreover, since the total pore volume is available for the low molecular weight contaminants, rather large volumes of sample may be desalted in one step and it is possible to yield only a small dilution. This method of desalting is faster and more efficient than dialysis and may be carried out both in small scale, e.g. a pipette-tip, and in larger liter scale.

Protein Fractionation

In preparative fractionation, the protein of interest is to be separated from other solutes of similar size and this separation put higher requirements on the choice of chromatography medium and selection of running conditions. From theoretical calculations it has been found that complete separation of molecules of spherical shape differing less than 30% in molecular mass may not be expected. This type of chromatography is popular as a polishing step for protein manufacture.

Determination of Molecular Size

The elution volumes of globular proteins are determined largely by their relative molecular size. Hence, the construction of a calibration curve, with proteins of a similar shape and known molecular weight, enables the molecular size values of other proteins to be estimated, even in crude preparations. In this way, SEC also provides a means of determining differences in the shape of native or denatured globular proteins under a wide variety of conditions.

In analytical SEC the resolution is of outmost interest and the correct choice of gel and operating conditions are critical if good results are to be obtained. Focus should thus be on selectivity rather than efficiency. To obtain maximum resolution, the starting zone must be narrow relative to the length of the column. The resolution of two separated zones in SEC increases as the square root of column length. In general the diameter of column is decided by the sample volume and the length by the resolution required.

Primary Structure Determination

Edman degradation is the process of purifying protein by sequentially removing one residue at a time from the amino end of a peptide. To solve the problem of damaging the protein by hydrolyzing conditions, Pehr Edman created a new way of labeling and cleaving the peptide. Edman thought of a way of removing only one residue at a time, which did not damage the overall sequencing. This was done by adding Phenyl isothiocyanate, which creates a phenylthiocarbamoyl derivative with the N-terminal. The N-terminal is then cleaved under less harsh acidic conditions, creating a cyclic compound of phenylthiohydantoin PTH-amino acid. This does not damage the protein and leaves two constituents of the peptide. This method can be repeated for the rest of the residues, separating one residue at a time.

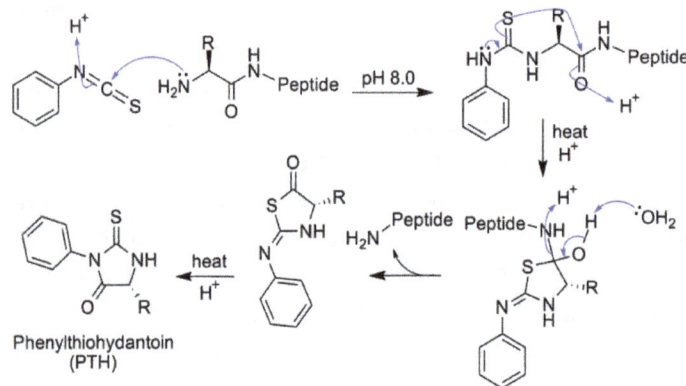

Edman degradation is very useful because it does not damage the protein. This allows sequencing of the protein to be done in less time. Edman sequencing is done best if the composition of the amino acid is known to determine the composition of the amino acid, the peptide must be hydrolyzed. This can be done by denaturing the protein and heating it and adding HCl for a long time. This causes the individual amino acids to be separated, and they can be separated by ion exchange chromatography. They are then dyed with ninhydrin and the amount of amino acid can be determined by the amount of optical absorbance. This way, the composition but not the sequence can be determined.

Sequencing Larger Proteins

Larger proteins cannot be sequenced by the Edman sequencing because of the less than perfect efficiency of the method. A strategy called divide and conquer successfully cleaves the larger protein into smaller, practical amino acids. This is done by using a certain chemical or enzyme which can cleave the protein at specific amino acid residues. The separated peptides can be isolated by chromatography. Then they can be sequenced using the Edman method, because of their smaller size.

In order to put together all the sequences of the different peptides, a method of overlapping peptides is used. The strategy of divide and conquer followed by Edman sequencing is used again a second time, but using a different enzyme or chemical to cleave it into different residues. This allows two different sets of amino acid sequences of the same protein, but at different points. By comparing these two sequences and examining for any overlap between the two, the sequence can be known for the original protein.

For example, trypsin can be used on the initial peptide to cleave it at the carboxyl side of arginine and lysine residues. Using trypsin to cleave the protein and sequencing them individually with Edman degradation will yield many different individual results. Although the sequence of each individual cleaved amino acid segment is known, the order is scrambled. Chymotrypsin, which cleaves on the carboxyl side of aromatic and other bulky nonpolar residues, can be used. The sequence of these segments overlap with those of the trypsin. They can be overlapped to find the original sequence of the initial protein. However, this method is limited in analyzing larger sized proteins (more than 100 amino acids) because of secondary hydrogen bond interference. Other weak intermolecular bonding such as hydrophobic interactions cannot be properly predicted. Only the linear sequence of a protein can be properly predicted assuming the sequence is small enough.

3D Structure Determination

Nuclear magnetic resonance, NMR, and X-ray crystallography are the only two methods that can be applied to the study of three-dimensional molecular structures of proteins at atomic resolution. NMR spectroscopy is the only method that allows the determination of three-dimensional structures of proteins molecules in the solution phase.

In addition NMR spectroscopy is a very useful method for the study of kinetic reactions and properties of proteins at the atomic level.

In contrast to most other methods NMR spectroscopy studies chemical properties by studying individual nuclei. This is the power of the methods but sometimes also the weakness.

NMR spectroscopy can be applied to structure determination by routine NMR techniques for proteins in the size range between 5 and 25 kDa. For many proteins in this size range structure determination is relatively easy, however there are many examples of structure determinations of proteins, which have failed due to problems of aggregation and dynamics and reduced solubility.

The Principle of NMR Spectroscopy

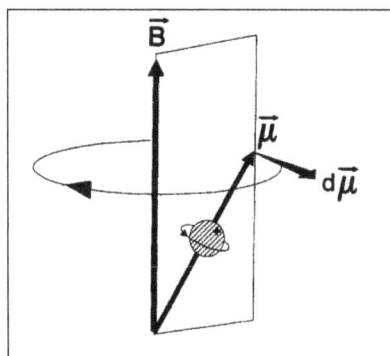

The spinning nucleus with a charge precessing in a magnetic field.

The atomic nuclei with odd mass numbers has the property spin, this means they rotate around a given axis. The nuclei with even numbers may or may not have this property. A spin angular momentum vector characterizes the spin. The nucleus with a spin is in other words a charged and spinning particle, which in essence is an electric current in a closed circuit, well known to produce a magnetic field. The magnetic field developed by the rotating nucleus is described by a nuclear magnetic moment vector, μ, which is proportional to the spin angular moment vector.

Larmor Precession

When a nucleus with a nuclear magnetic moment is placed in an external magnetic field B, Figure, the magnetic field of the nuclei will not simply be oriented opposite to the orientation of the magnetic field. Because the nucleus is rotating, the nuclear magnetic field will instead precess around the axis of the external field vector. This is called Larmor precession. The frequency of this precession is a physical property of the nucleus with a spin, and it is proportional to the strength of the external magnetic field, the higher the external magnetic field the higher the frequency.

Making Larmor Precession of a Nuclear Spin Observable

The magnetic moment of a single nucleus cannot be observe. In a sample with many nuclei the magnetic moment of the individual nuclei will add up to one component. In a sample of nuclei in a magnetic field the component of the magnetic moment of the precessing nuclei around the external field axis will be a vector in the direction of the field axis. This nuclear magnetization is impossible to observe directly. In order to observe the nuclear magnetization we want to bring the nuclear magnetization perpendicular to the applied field. Applying a radio frequency pulse, which is perpendicular to the external magnetic field, can do this. If this pulse has the same requency as the Larmor frequency of the nuclei to be observed and a well defined length the component of the nuclear magnetization can be directed from the direction of the magnetic field to a direction perpendicular to this. After the pulse the nuclear magnetization vector will be rotating in the plane perpendicular to the magnetic field. The angular rate of this rotation is the Larmour frequency of the nuclei being observed. The rotating magnetic field will induce an electric current, which can be measured in a circuit, the receiver coil, which is placed around the sample in the magnet.

The Free Induction Decay, FID

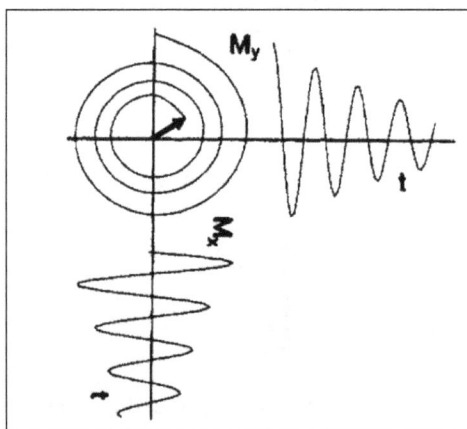

The free induction decay, FID, as measured as a function of time
in the x- and y-directions perpendicular to magnetic field.

The nuclear magnetization perpendicular to magnetic field will decrease with time, partly because the nuclear magnetization gets out of phase and because the magnetization returns to the direction of the external field. This will be measured in the receiver coil as a fluctuating declining amplitude with time. This measures a frequency and a decay rate as a function of time. This is the direct result of the measurement and referred to as the free induction decay, FID.

Fourier Analysis

In order to get from the FID to an NMR spectrum the data in the FID are subjected to Fourier analysis. The FID is a function of time; the Fourier transformation converts this to a function of frequency. This is the way the NMR spectrum is normally displayed.

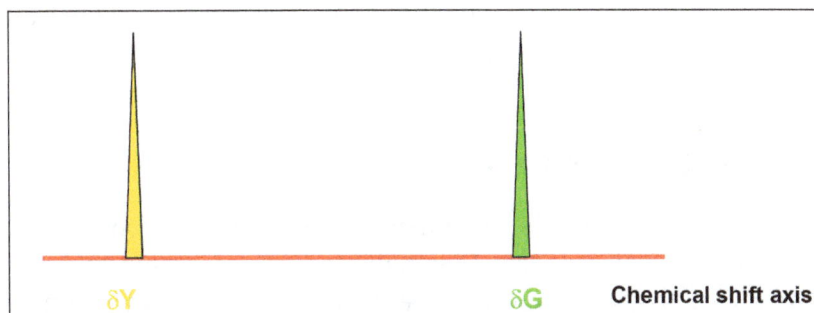

Schematic presentation of an NMR spectrum of a compound
with two nuclei of different chemical shift.

The Origin of Chemical Shift

When a molecule is placed in an external magnetic field, this will induce the molecular electrons to produce local currents. These currents will produce an alternative field, which opposes the external magnetic field. The total effective magnetic field that acts on the nuclear magnetic moment will therefore be reduced depending on strength of the locally induced magnetic field. This is called a screening effect, a shielding effect, or more commonly known as the chemical shift. The shielding from the external magnetic field depends on the strength of the external magnetic field, on the chemical structure and the structural geometry of the molecule. The NMR active nuclei of the molecule, therefore, experience slightly different external magnetic fields, and for this reason the resonance conditions are slightly different. The precession of the nuclear magnetic moment vector of the individual nuclei will have different angular velocities. The FID of a molecule will therefore have several frequency components, and the Fourier transformation will produce an NMR spectrum with signals from each of the different types of nuclei in the molecule.

Measuring the Relative Chemical Shift

The chemical shift for an NMR signal is normally measured in Hz shifted relative to a reference signal. In order to compare NMR spectra obtained at different external magnetic fields the difference in chemical shift is divided by the spectrometer frequency, which is the Larmour frequency of the observed nucleus type for the strength of the magnetic field of the spectrometer magnet. The chemical shift is normally a very small number in Hz, divided by a very large number the spectrometer frequency also in Hz. Therefore the resulting small number is multiplied by one million and given in the dimension less unit of parts per million.

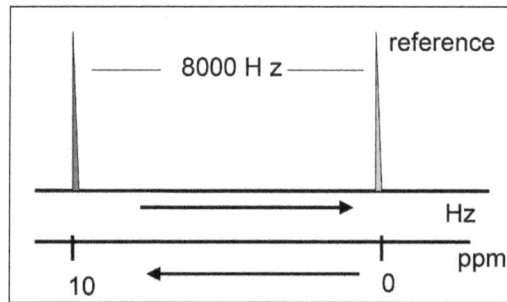
Schematic presentation of the chemical shift axis. The top shows an NMR spectrum with two signals.

One signal (right) is from a chemical compound, which has been added to the sample for reference. The other signal is from a nucleus in the compound of interest. The two signals are 8000 Hz apart and the spectrometer frequency is 800 MHz. The two signals are $8000*10^6 /800*10^6$ ppm = 10 ppm apart. If the reference is set to 0 ppm the signal of interest is at 10 ppm, bottom axis.

Increasing Field and Decreasing ppm Values

In contrast to the conventional axis, the ppm axis in NMR spectroscopy is a left oriented axis, counting increasing numbers towards left. Originally the axis was used to describe the effective magnetic field at the nuclei observed. In this respect the axis is a fully conventional right oriented axis, with increasing effective field towards right.

NMR of the 20 Common Amino Acids

The ^{1}H (proton) NMR summary of the 20 common amino acids is shown in Table as a listing of the chemical shifts for the hydrogen atoms of the residues in a random coil peptide.

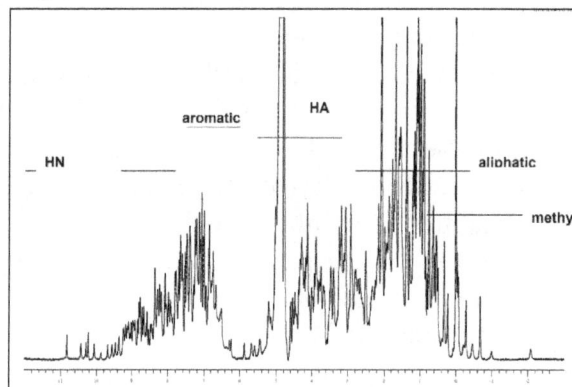
^{1}H-NMR spectrum of hen egg white lysozyme.

The NMR Spectrum of a Folded Protein

In a protein structure the individual residues are packed into chemical environments, which are very different from the random coil situation. This is seen in Figure that shows the NMR spectrum of a folded protein and a comparison with the NMR spectrum of the same protein recorded at conditions where the protein is unfolded. The spectrum of the unfolded protein corresponds to a spectrum, which in essence is the sum of the random coil spectra of the amino acid residues in the proteins as given in Table. The dispersion of signals in the spectrum of the folded protein is far beyond the envelope of signals seen in the spectrum of the unfolded protein. This clearly reflects that nuclei in the folded form are subject to a many different types of microenvironments of chemical screens.

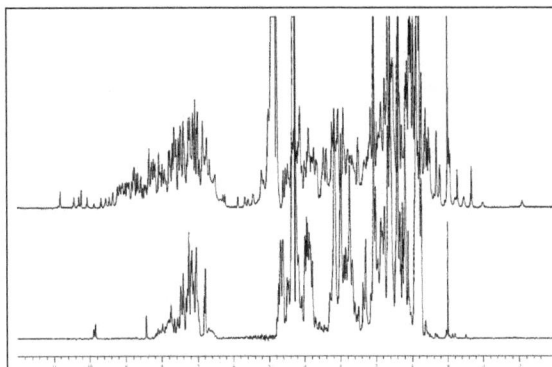

Comparison of NMR spectra of folded (top) and unfolded (bottom) protein.

Small and Large Deviation from Random Coil Shifts

For NMR spectra of proteins the NMR signals of the nuclei of the individual residues are in most cases seen in the vicinity of the random coil shift value. However, due to structural diversity in a protein structure, the signals are often shifted significantly away from the random coil value. In some cases, a proton may be in a chemical environment of a protein structure, where the chemical shift of the external magnetic field is so strong that the corresponding signals are observed several ppms away from the random coil values. The signal observed at −2 ppm in the spectrum of lysozyme is from a one of the gamma methylene protons of isoleucine 98. This is shifted more than 3 ppm upfield from the random coil position at 1.48 ppm due to the proximity of a tryptophan residue.

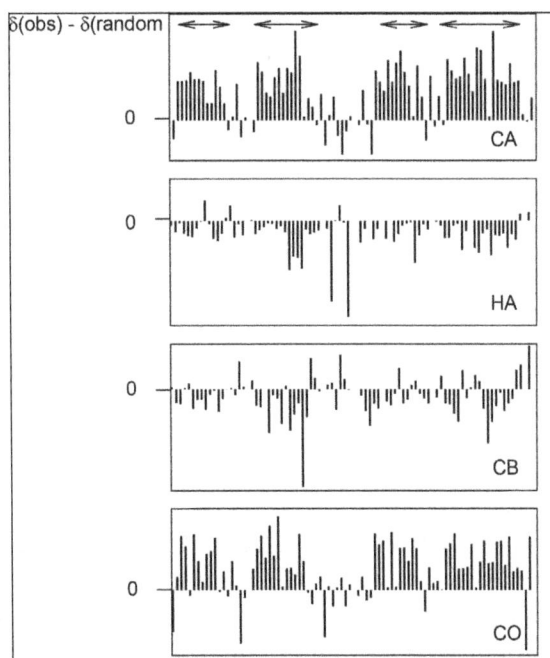

Chemical shift analysis of the peptide backbone NMR signals. The protein studied has four α-helices as marked by the arrows in the top panel. Each panel represent a chemical shift analysis of the individual residues comparing the observed shift wit the shift observed for the same residue type in random coil model peptides.

Residue	NH	αH	βH	Others
Gly	8.39	3.97		
Ala	8.25	4.35	1.39	
Val	8.44	4.18	2.13	γCH_3 0.97, 0.94
Ile	8.19	4.23	1.90	γCH_2 1.48, 1.19
				γCH_3 0.95
				δCH_3 0.89
Leu	8.42	4.38	1.65,1.65	γH 1.64
				δCH_3 0.94, 0.90
Pro [b]		4.44	2.28,2.02	γCH_2 2.03, 2.03
				δCH_2 3.68, 3.65
Ser	8.38	4.50	3.88,3.88	
Thr	8.24	4.35	4.22	γCH_3 1.23
Asp	8.41	4.76	2.84,2.75	
Glu	8.37	4.29	2.09,1.97	γCH_2 2.31, 2.28
Lys	8.41	4.36	1.85,1.76	γCH_2 1.45, 1.45
				δCH_2 1.70, 1.70
				εCH_2 3.02, 3.02
				εNH_3^+ 7.52
Arg	8.27	4.38	1.89,1.79	γCH_2 1.70, 1.70
				δCH_2 3.32, 3.32
				NH 7.17, 6.62
Asn	8.75	4.75	2.83,2.75	γNH_2 7.59, 6.91
Gln	8.41	4.37	2.13,2.01	γCH_2 2.38, 2.38
				δNH_2 6.87, 7.59
Met	8.42	4.52	2.15,2.01	γCH_2 2.64, 2.64
				εCH_3 2.13
Cys	8.31	4.69	3.28,2.96	
Trp	8.09	4.70	3.32,3.19	2H 7.24
				4H 7.65
				5H 7.17
				6H 7.24
				7H 7.50
				NH 10.22
Phe	8.23	4.66	3.22,2.99	2,6H 7.30
				3,5H 7.39
				4H 7.34
Tyr	8.18	4.60	3.13,2.92	2,6H 7.15
				3,5H 6.86
His	8.41	4.63	3.26,3.20	2H 8.12
				4H 7.14

Random chemical shift of the 20 common amino acid residues in proteins.

Determination of secondary structure from chemical shift analysis The chemical shift can be used in structure determination of proteins. In particular the analysis of the chemical shift of $^1H^\alpha$, and the ^{13}CO, $^{13}C^\alpha$, and $^{13}C^\beta$ backbone can be used to determine the secondary structure type of a given peptide segment.

Multidimensional NMR Spectroscopy

The fundamental NMR spectrum is a one-dimensional frequency spectrum, The FID of one-dimensional NMR spectrum represents only one time dimension. In two-dimensional NMR spectroscopy the second dimension might be another time domain, which can be established by introducing an additional pulse and systematically increasing the time between the two pulses, And equivalently for the three-dimensional NMR spectrum. This can be combined with the observation of one or two more nuclei for instance ^{13}C and ^{15}N.

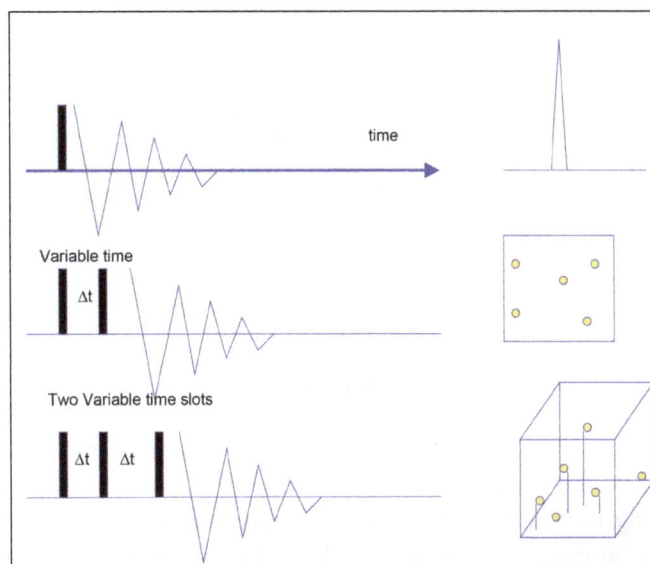

Schematic presentation of NMR spectra in one-dimension, top, twodimensions, middle, and three-dimensions.

J Coupling Origin and Application in Structure Determination

J coupling is also referred to as spin-spin coupling and scalar coupling. Two nuclei in a molecule, which are connected by one, two, and three bonds, can be seen to be coupled in the NMR spectrum. The coupling is observed by a splitting of the NMR signal. The origin of this coupling is a process in which the two nuclei perturb the respective valence electrons of the molecule. Here the electron spin and the nuclear spin at one atom are being aligned anti-parallel to each other and as a response to this the electrons of the coupling nucleus orients either parallel or anti-parallel to that of A and vice versa. For two nuclei which can each exist in two states there will be four different types of interactions, which is reflected in the A and B signals are both splitting into two signals.

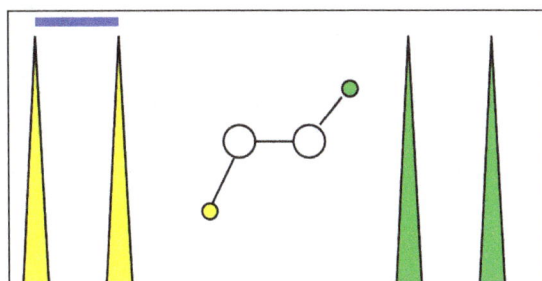

J- coupling of two nuclei separated by three bonds of compound of the type Y-C-C-G. The blue bar shows the size of the coupling.

The coupling constant J

The separation between the two components of the split signal is the coupling, This is normally measured in Hz and referred to as the coupling constant. The coupling constant is independent of the magnetic field. The size of the coupling depends on many structural properties. For structure determination we are mainly concerned with the relationship between the coupling constant and dihedral angles.

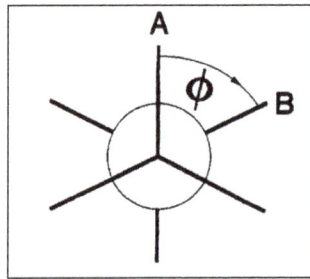

The definition of a dihedral angle φ as seen in Newman projection.

J-couplings and Determination of Dihedral Angles

The three-bond coupling constant depends on the dihedral angle defined by rotation around the middle bond in the coupling system. The definition of the dihedral angle is shown in the Newman projection. The J-coupling may also be used to distinguish between trans and gauche conformations. One particularly important application of the coupling constant is as a measure of the coupling between the Hα and the HN in the peptide backbone. This coupling depends on the φ- angle in the peptide bond. The coupling may also be measured by he coupling constant between the HN and Cβ. The Karplus curve shown in figure shows the correlation between the coupling between Hα and the HN and the φ- angle. It is seen that coupling constants is around 4 Hz for peptide segments in α helices where the φ- angle is around -60°, and it is between 8 and 12 Hz for peptide segments in β-structures, where the φ- angle is in the -120° range.

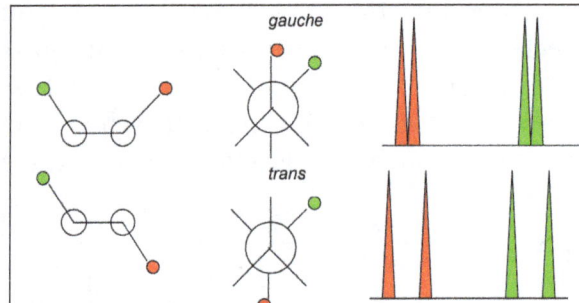

The coupling constant can distinguish between the gauche and trans conformation.

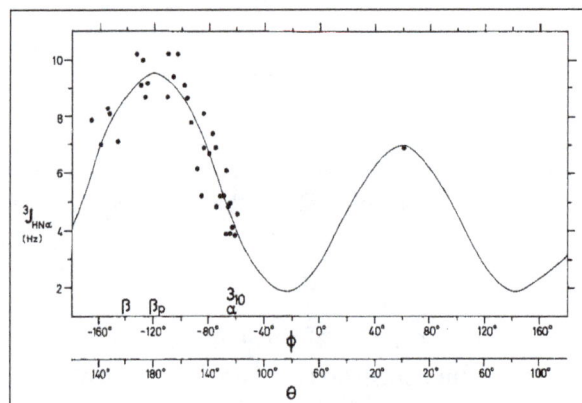

The Karplus function shows the correlation between the φ-angle in the peptide bond and coupling constant between Hα and HN.

- The use of J-coupling for Identification of the Amino Acid Residues in the NMR Spectra of Proteins.

The One Dimensional ^{1}H NMR Spectrum of Valine

Most of the methods used in NMR spectroscopy are designed to measure the presence of coupling between nuclear spins. If we consider the amino acid residue of valine in a peptide this will have NMR signals from the H^N, the H^α, the H^β, and the two triple intensity signals of $H^{\gamma1}$ and $H^{\gamma2}$. The schematic ^{1}H NMR spectrum of a valine residue is shown in figure, where δH^N is the chemical shift of the HN signal etcetera. This one-dimensional spectrum shows the individual signals of the ^{1}H nuclei, however, it does not provide information about the spin-spin coupling partners of the five corresponding nuclei. This information can be obtained by two-dimensional correlation NMR spectroscopy, COSY.

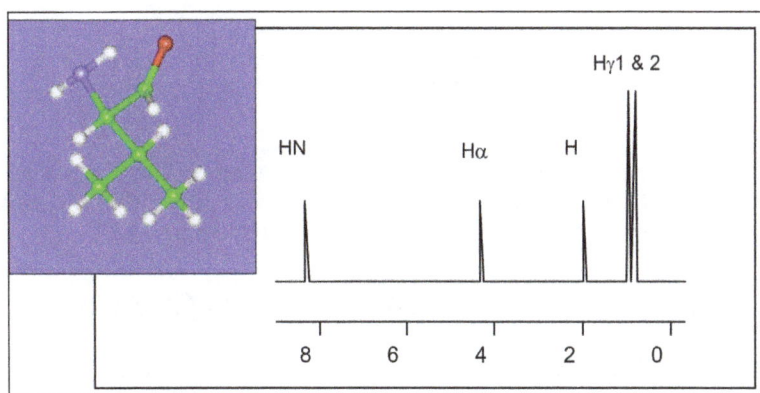

Schematic presentation of the NMR spectrum of Valine.

The COSY Spectrum of Valine

From the covalent structure of the valine residue it is seen that the HN spin couples to the Hα spin, which couples to the Hβ spin which couples to the two sets of H$^\gamma$ spins. The two-dimensional COSY spectrum, figure, is recorded so that the spectrum contains two types of signals, diagonal peaks and off-diagonal peaks often called cross peaks. The diagonal peaks represent the signals from each the of the ^{1}H types in valine (H^N, H^α, H^β, and the two H^γ) and their position in spectrum are at (δHN, δH^N), (δH^α, δH^α), (δH^β, δH^β), ($\delta H^{\gamma1}$, $\delta H^{\gamma1}$) and ($\delta H^{\gamma2}$, $\delta H^{\gamma2}$). The cross peaks are, however, the most interesting, they report the couplings between pairs of nuclei. The coupling between HN and Hα, show at the two symmetry related positions (δH^N, δH^α) and (δH^α, δH^N), figure. Similarly the coupling between Hα and Hβ are at (δH^α, δH^β) and (δH^β, δH^α), and the couplings between H$^\beta$, and the H$^{\gamma1}$ and H$^{\gamma2}$, respectively, at (δH^β, $\delta H^{\gamma1}$) and ($\delta H^{\gamma1}$, δH^β), and (δH^β, $\delta H^{\gamma2}$) and ($\delta H^{\gamma2}$, δH^β).

The TOCSY Spectrum of Valine

By recording the correlation NMR spectrum in a way so that all the spins in a spin system of an amino acid are all correlated results in a more complicated spectrum.

However this type of spectrum becomes very useful in particular in the heteronuclear NMR experiments a comparison of the TOCSY and COSY spectrum can be seen in figure below.

COSY and TOCSY Spectra of the Amino Acid Residues in Proteins

Many of the types of residues of the amino acids have characteristic patterns of couplings in combinations with characteristic chemical shifts, which make these unique for identification. Other residue types are quite similar and not so easy to identify just on the basis of their coupling patterns and chemical shifts. The COSY and TOCSY pattern for each of the individual amino acid residues can easily be drawn schematically from the information in table.

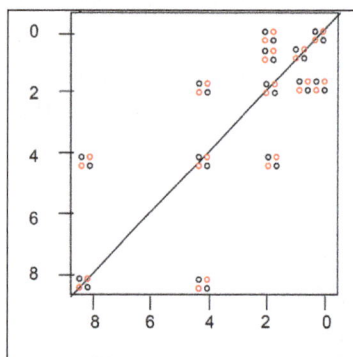

Schematic presentation of a COSY spectrum of the amino acid residue valine. The pattern of four circles represents on peak. The peaks on the diagonal are the diagonal peaks. The off-diagonal peaks are the cross peaks. Try to assign the spectrum.

Assigning the Amino Acid Residue Spin System by COSY and TOCSY

The assignment of the ^{1}H NMR spectrum of a protein will normally use a starting point in the coupling between HN and Hα of a residue often appearing as well resolved signal in the (δH^N, δH^α) and (δH^α, δH^N) region of the ^{1}H NMR spectrum. The assignment may then continue to identify the remaining signals of the residue. On the basis of the number of coupling spins and their chemical shifts it is possible to assign the systems of spins to an amino acid type or to a group of similar amino acid types, which have similar number of spins with similar chemical shifts.

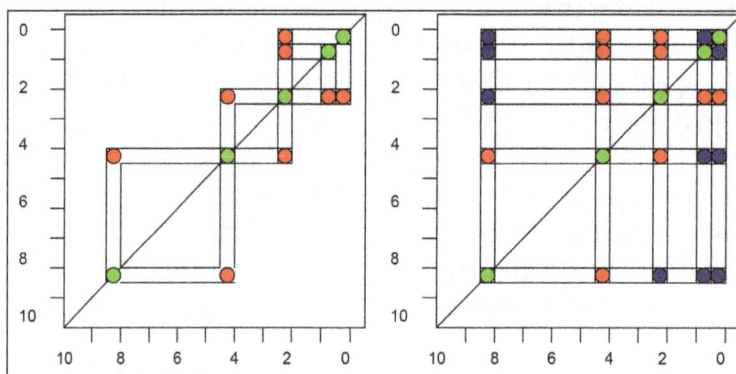

Schematic presentation of COSY (left) and TOCSY (right) spectra of a valine residue. Green circles are diagonal peaks, red circles are two. And threebond couplings seen only in the COSY spectrum, blue circles are TOCSY peaks representing long range coupling in the spin system.

Introducing NMR Active Stable Isotopes ^{13}C and ^{15}N

It is often a great advantage for the analysis of a protein by NMR to introduce the NMR active stable isotopes ^{13}C and ^{15}N. With the introduction of these two NMR active nuclei the spins in a protein are almost all being connected by one-bond couplings, and this facilitates the study tremendously. The preparation of proteins enriched with the two nuclei are accomplished by heterologous expression of the protein in micro-organisms grown in minimal growth medium where the carbon source is fully ^{13}C labelled and the nitrogen source is fully ^{15}N labelled.

- The ^{1}H-^{15}N Coupling in the Peptide Bond is the Starting Point for the Heternuclear NMR Analysis of Proteins.

The one-bond coupling ^{1}H-^{15}N is the most important starting point for the heternuclear NMR analysis of proteins. This bond is present in every amino acid residue in a protein except the N-terminal and the proline residues. The correlation spectroscopy method used to record this coupling is called a ^{1}H-^{15}N HSQC spectrum (heteronuclear single quantum correlation). An example of this spectrum is shown in figure. In this spectrum there is a ^{1}H-chemical shift axis and a ^{15}N chemical shift axis. The signals report the coupling between the HN and N and the signal appears at (δH^N, δN).

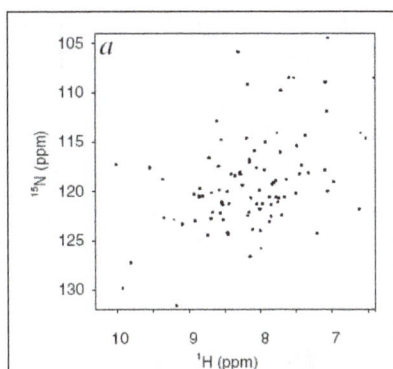

^{1}H-^{15}N HSQC spectrum of the ^{15}N labelled protein ACBP. Each "spot" is an NMR signal representing the ^{1}H-^{15}N coupling form one of the amino acid residues in the protein.

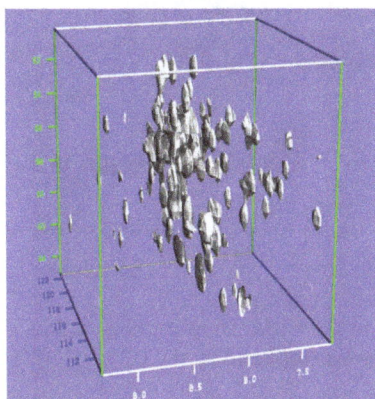

Tripleresonance heteronuclear NMR spectrum of a protein. Each signal represents the coupling between H^N and N with C$^\alpha$ in one residue. The projection of the signals into the ^{1}HN and ^{15}N plane will show a spectrum.

With three NMR active nuclei fully incorporated in a protein experiments in three dimensions are possible, one for each type of isotope. This is a heteronuclear tripleresonance experiment. The list of experiments that can be used to identify couplings between the three nuclei in amino acid residues are many.

Above in figure is shown a triple resonance heteronuclear correlation spectrum, which correlates the coupling between H^N and N with C^α in one residue. The three-dimensional spectrum has three chemical shift axes a ^{1}H-axis, a ^{13}C-axis and a ^{15}N-axis, and the signals appear at (δH^N, δ^N, δC^α).

Sequential Assignment using Heteronuclear Scalar Coupling

Sequential assignment is a process by which a particular amino acid spin system identified in the spectrum is assigned to a particular residue in the amino acid sequence. In a protein, which is ^{13}C- and ^{15}N labelled almost the entire protein is one continuous spin system. In particular the peptide backbone is one long series of scalar coupled nuclei. Specific types of heteronuclear correlation spectroscopy can record these individual types of coupling. There are several principles for sequential assignments using heteronuclear scalar coupling. One principle is illustrated in figure below. It is based on the recording of two different heteronuclear correlation spectra. One ^{1}H-, ^{13}C-, and ^{15}N-heteronuclear three-dimensional NMR spectrum, which records the one bond coupling between ^{1}HN and ^{15}N and the one and two bond coupling between ^{15}N and ^{13}C$^\alpha$ and ^{13}C$^\beta$ in one residue. The spectrum is called a HNCACB spectrum. This type of experiment also records the coupling across ^{13}C' to the ^{13}C$^\alpha$ and ^{13}C$^\beta$ in the preceding residue. Another similar experiment measures specifically the heteronuclear coupling between ^{1}HN and ^{15}N in one residue and the coupling across ^{13}C' to the ^{13}Cα and ^{13}C$^\beta$ in the preceding residue. This spectrum is called a CBCA (CO)NH spectrum. In a combined analysis of these two types of spectra it is possible from each individual ^{1}HN - ^{15}N peak in the ^{1}HN - ^{15}N correlation spectrum, Figure to identify the ^{13}C$^\alpha$ and ^{13}C$^\beta$ in the same residue and the preceding residue. The principle in the method is demonstrated in figure below.

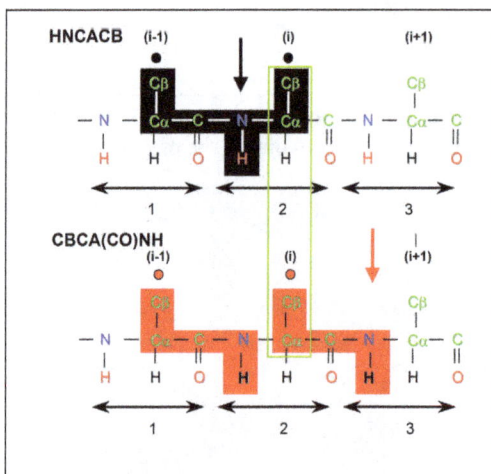

The two panels show the scalar coupling correlation, which is measured by the HNCACB (top) and by the CBCA(CO)NH (bottom).

- In the HNCACB (top) the coupling is mediated through the chemical bonds shown on a black background. The ^{1}H^N –^{15}N coupling pair of residue (i) is correlated to the ^{13}C$^\alpha$ - ^{13}C$^\beta$ pair of residue (i) and (i-1).

- In the CBCA(CO)NH (bottom) the coupling is mediated through the bonds shown on a red background. Here the ^{1}H^N –^{15}N coupling pair of residue (i) is correlated to the ^{13}C$^\alpha$ - ^{13}C$^\beta$ in residue (i-1).

- If the same ^{13}C$^\alpha$ - ^{13}C$^\beta$ pair, as shown in the open green frame, are seen to couple to two different pairs of $^1H^N$ –^{15}N couplings as indicated by the black and red arrows in the two panels, they may be assigned as signals from neighbouring residues.

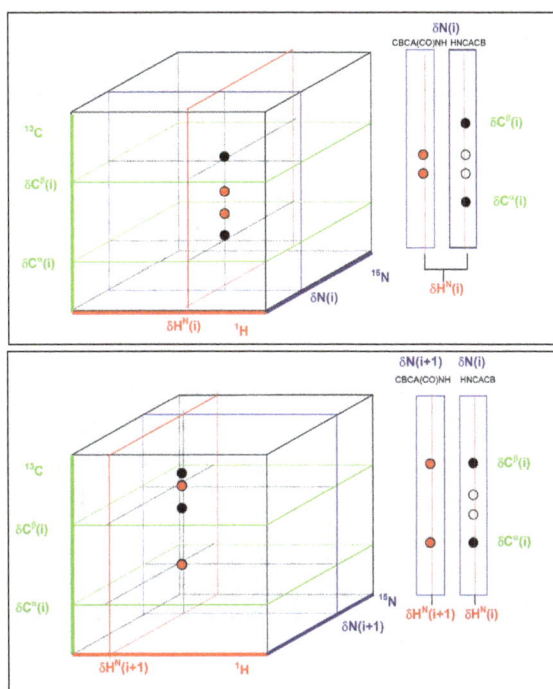

Schematic presentation of the combined spectrum analysis of the threedimensional HNCACB and a CBCA(CO)NH pectra. The 15N axis and the frames in the 15N dimension are coloured blue. The1 H axis and the frames in the 1 H dimension are coloured red. The 13C axis and the frames in the 13C dimension are green. Two planes δN(i), top-left, and δN(i+1), bottom-left, are highlighted.

The cross peaks in HNCACB are • and in CBCA(CO)NH are:

- To the right are shown the segments of the planes from the two types of spectra. In the top-right the HNCACB and the CBCA(CO)NH from the same plane, δN(i), of the two spectra are compared.

- In the bottom-right δN(i) plane from the HNCACB is compared to the δN(i+1) planes of the CBCA(CO)NH.

- The δN(i) plane of the CBCA(CO)NH spectra has two signals at the δH(i). These are the red cross peaks and origins from C^β and C^α of residue (i-1).

- The δN(i) plane of the HNCACB has four cross peaks (black) at the δH(i). Two of these superimpose with the red cross peaks of the CBCA(CO)NH are from the preceding residue (i-1). The observation establishes a sequential assignment.

- The δN(i) plane of the HNCACB and the δN(i+1) plane of the CBCA(CO)NH have one pair of C^β and C^α cross peaks in common. The observation establishes a sequential assignment.

As described in figure above the sequential assignment brings groups of spin systems together in a sequence. This sequence can be imbedded in the amino acid sequence of the protein by a chemical shift analysis of the spin systems. Several of the residue types have typical chemical shifts and this can be used for a residue assignment. For instance a series of sequentially assigned spin systems the chemical shift analysis identifies:

GXX(T/S)XXXAXXGXX.

Which can be imbedded between residue 132 and 144 in the amino acid sequence:

132 GNRSKDVAIVGLL 144.

Leading to the assignment of the intervening residues, which were not assigned directly by the chemical shift analysis.

Dipole-dipole Interactions

A nuclear spin is a magnetic dipole with a magnetic field. In a hypothetic situation in the presence of an external field the total magnetic field at the spin will be the sum of the local field and the external field. If two identical nuclear spins are close to each other in an external magnetic field, the two nuclear spins will be influenced by the external field and by the magnetic field of the other nuclear spin. The local field may be either aligned or parallel to the external field, and this will give rise to two signals. The separation between the two signals depends on the distance between the two spins and the angle between directions of the vector connecting the two spins and the vector representing the external magnetic field.

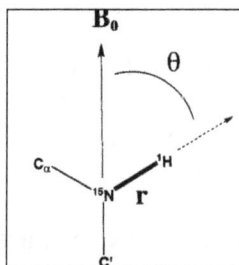

The dipole-dipole coupling between the ^{15}N and the ^{1}H of the NH one bond vector of the peptide backbone depends on the length of the bond and the angle θ between the direction of the bond vector and the direction of the external magnetic field vector.

In samples of solid material where nuclear spins are heterogeneously maintained in fixed positions relative to each other this gives rise to extremely broad lines. In solution where there is molecular motion the dipole-dipole interactions are averaged away and not observed.

Residual Dipolar Coupling

It is possible to re-establish the dipolar coupling in solutions of charged colloid particles. In very strong magnetic fields it is possible to orient homogenously charged particles. Protein molecules placed in such environments will no longer have free molecular motion and the dipole-dipole interactions will be partly re-established. The degree to which the residual dipolar coupling is re-established can be controlled by the concentration of colloidal solution. The reason for being interested in reestablishing the dipolar coupling is the angular dependence to the magnetic field, figure. In the isotope enriched protein molecule there are several dipole-dipole interactions between spins of fixed distances maintained in single bonds, for instance the HN-N bond vector of the peptide backbone, figure below. By measuring residual dipolar coupling for spin-pairs with fixed distances it is possible to relate all the angles of these bonds to the direction vector of the magnetic field, and subsequently to each other. The use of residual dipolar coupling has in many cases been proven to have an enormous effect on the accuracy of the structure determination by NMR spectroscopy.

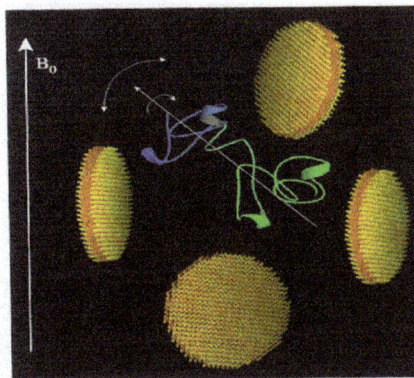

Oriented bicelles in a magnetic field. The protein molecule dissolved in the colloidal solution has anisotropic motion, which reintroduces dipolar coupling.

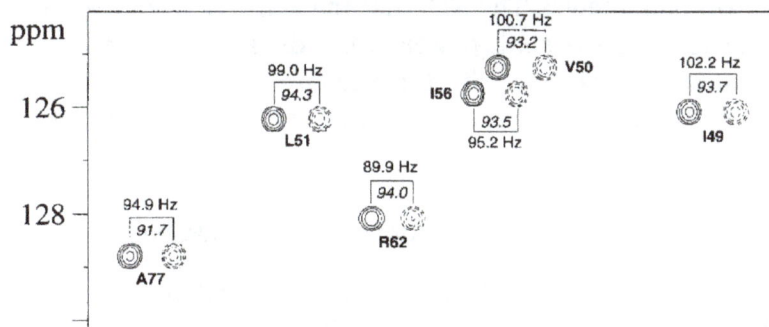

An example of measuring a residual dipolar coupling. The one bond coupling between ^1H and ^{15}N in the peptide bond has been measured in a magnetically oriented colloidal solution and in the freely tumbling protein. The coupling is the sum of the J-coupling and the residual dipolar coupling.

Cross Relaxation and the Nuclear Overhauser Effect

Cross relaxation is a result of dipole-dipole interaction between proximate nuclear spins. When two spins are very close in space, they experience each other's magnetic dipole moment. It is possible by NMR pulse techniques to either reverse the direction of the magnetic dipole of the nuclear spin or to "turn off the magnetic dipole" of one of the spins and to measure, how this affects the other spin. The rate by which this effect is transmitted to the other nuclear spin is the cross relaxation rate, and this is inversely proportional to the sixth power of the distance between the two nuclear spins. The cross relaxation rate and the related nuclear Overhauser effect can be used to estimate distances between two nuclei in a protein molecule. The effect is not only depending on the distance between the two nuclei it also depends on the overall molecular motion and on internal motions in the protein, which change the distance between the two nuclei with time.

The nuclear Overhauser effect is one of the most important tools in structure determination by

NMR spectroscopy. The experiment used to determine the nuclear Overhauser effect is called NO-ESY derived from Nuclear Overhauser Effect Spectrosco. The most common experiment is a homo-nuclear two-dimensional NMR experiment. In figure is shown a schematic presentation of a twodimensional NOESY spectrum between two nuclei R and G.

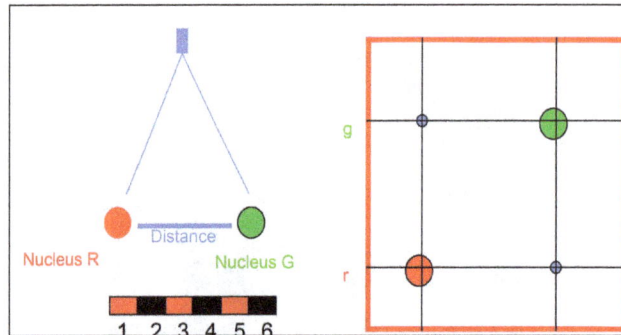

Schematic presentation of a two-dimensional NMR spectrum recording the Nuclear Overhauser effect between nucleus R and nucleus G. The effect is measured by the intensity of the blue cross peaks at (r,g) and (g,r).

The nuclear Overhauser effect can typically be measured between nuclei, which are less the 0.5 nanometers (10-9 meter) apart. Many structural chemists prefer to use Ångström as the unit of length (1 Ångström = 0.1 nanometer). The NOE decline with the distance between the two nuclei being inversely proportional to the sixth power of the distance between the two nuclear spins. This means that if a nuclear Overhauser effect is measured to be one for two nuclei, which are 5 Ångstrøms apart, the NOE between two nuclei, which are 2 Ångstrøms apart, will be $(5/2)^6 \approx 244$ times larger.

Sequential Assignment using Homonuclear ^{1}H NMR NOE Spectroscopy

It is not always feasible to produce a ^{13}C, ^{15}N double labelled protein sample. In this case samples may either be studied by ^{1}H- NMR or by a combination of ^{15}N and ^{1}H NMR spectroscopy. Here the nuclear Overhauser effect can be used to make sequential assignments. In the two most common types of secondary structure the peptide chain accommodates conformation which bring ^{1}H of the peptide backbone and the side chains of neighbouring residues so close together that they are observable by NOE spectroscopy. In the α-helix the neighbouring HN are 2.8 Ångström apart, and in the β-strand the distance from Hα in residue (i) to HN in residue (i+1) is only 2.2 Ångström.

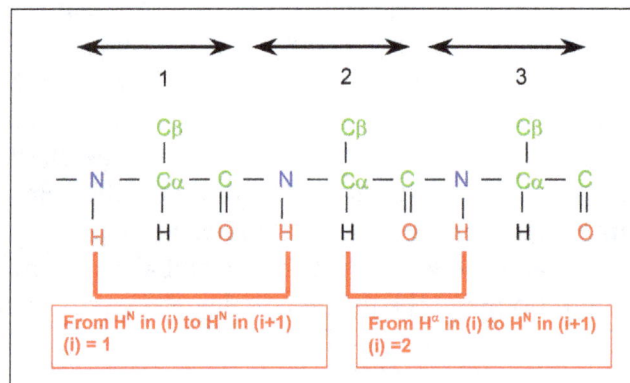

Two important sequential assignment tools using sequential assignment by NOEs.

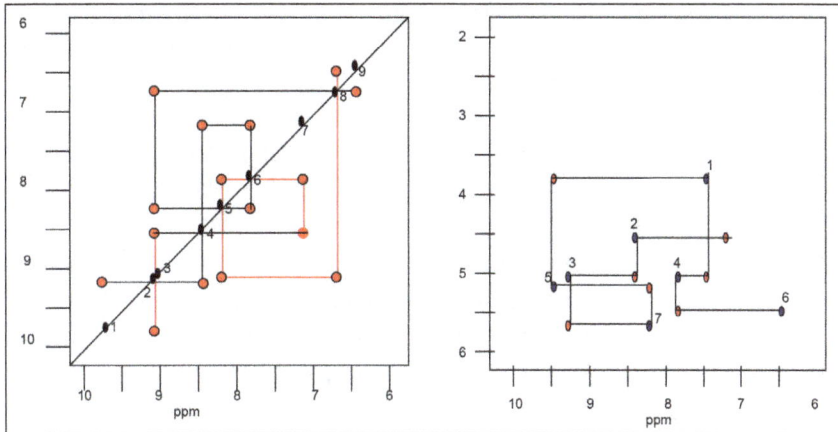

In figure above left, Schematic presentation of a NOESY spectrum in the region where nuclear Overhauser effects between H^N atoms are observed. The diagonal cross peaks have been marked as black signals and labelled 1 to 9. The red cross peaks are sequential NOEs.

In figure above right, Schematic presentation of sequential assignment of by the NOE between H^α in residue (i) to HN in residue (i+1). The schematic presentation shows the superimposition of a COSY spectrum with blue annotated cross peaks and a NOESY spectrum with red cross peaks. The blue cross peaks are the scalar (through bond) coupling between $H\alpha$ and HN in one residue. The red cross peaks are the sequential NOE between $H\alpha$ in residue (i) to HN in residue (i+1).

The red and black lines in figure below left show the sequential assignment of nine HN residues of a helix, where the order of sequence of the HN atoms in the peptide chain is revealed by the NOE connectivity of the signals in the order 1, 2, 4, 7, 6, 5, 3, 8, 9 or the reverse.

If it is known that signal 2 is from a valine, 5 from a tryptophan and 9 from a glycine the sequential assignment may be performed by matching the sequences XVXXXWXXG or the reverse GXX-WXXXXVX to the known amino acid sequence of the protein, so a fit would be the sequence:

- 983567421 sequence of signals.

- GAKWSRYVP amino acid sequence.

Using the signal annotation the sequential in figure below right the assignment goes 6, 4, 1, 5, 7, 3 2. Here the direction of the sequence is unambiguous because the direction is always from H^α in residue (i) to H^N in residue (i+1). As above the signal sequence can be imbedded in the amino acid sequence if two or three of the spin systems of the connected signals have been assigned.

Structure Determination by NMR Spectroscopy

The structure determination by NMR spectroscopy depends critically on the assignment of the NMR signals of the NMR active nuclei of the protein. NMR spectroscopy has information about the structural geometry around every NMR observable nuclei in the protein. This information may be read from the NMR signals of the nuclei. It is therefore important that the signals in the NMR spectrum have been assigned to the correct nuclei, those, which give raise to the signals. From the NMR signals the distance to the near-by nuclei in the structure can be read using NOE spectroscopy, the dihedral angles can be determined by coupling constants and chemical shifts, the relative angles of a number of bond vectors can be determined by residual dipolar coupling.

NMR spectroscopy provides several sets of information, which are all being used for the structure calculation. The computer programs, which handles the structure calculation is a molecular dynamics program. As a starting point the program has information about the covalent structure geometry of all the common amino acids regarding bond lengths and angles and potential functions to ensure that the bonds are maintained within realistic limits. For non-defined distances the program has a set of potentials that controls electrostatic interactions and van der Waals interactions ensuring that atoms do not get closer to each other than the Lennard-Jones potential permits.

The structural information from NMR is typically entered into the program either as a distance range or an angular range. The computer uses potential functions to ensure that every single NMR derived input structure data stay in the set range. The molecular dynamics program is using a process called simulated annealing where the protein atoms are heated and cooled successively, while the potential functions are turned on to form the structure. The repeated heating and cooling process is meant to help energetically unfavourable structures to overcome energy barriers and end up in energetically more favourable structures, which may resist the subsequent heating process.

Because the NMR derived structural input cannot provide a unique structure input for all the atoms in the protein, the NMR data set cannot define the structure unambiguously. Therefore, the structure calculation based on NMR derived structure constraints takes its origin in a randomly generated structure. The structure determination is repeated several times with new and different starting points. Typically one hundred structures are calculated, and those structures, which comply best to the NMR input data and are energetically most favourable, are selected as group of structures often referred to as an NMR bundle. The bundle is the result of a structure determination by NMR, figure below. The superimposed structures reveal how reproducibly the structure calculation program calculates the structure. This is measured as the root mean square deviation, RMSD, of every single atom in the structure. The rmsd for the peptide backbone atoms is the sum of the RMSDs for all the atoms in the peptide backbone divide by the number of atoms.

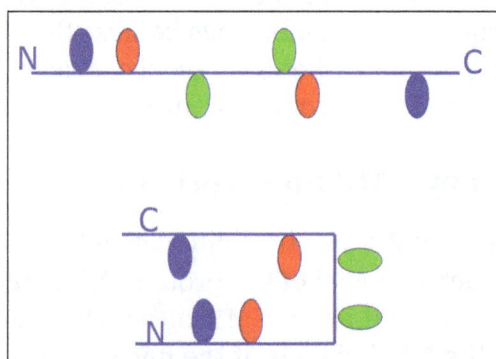

Schematic demonstration of using distance constraints to calculate a structure. In the upper figure is shown a linear strand with beads. The green, red and blue pairs of beads, respectively have been shown to be close to each other. The structure below represents one solution to determining the structure based on the three pieces of distance information.

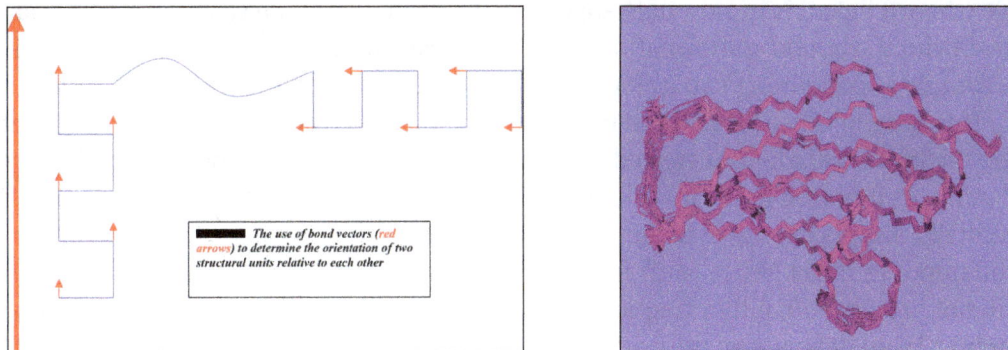

NMR bundle, 20 structures superimposed.

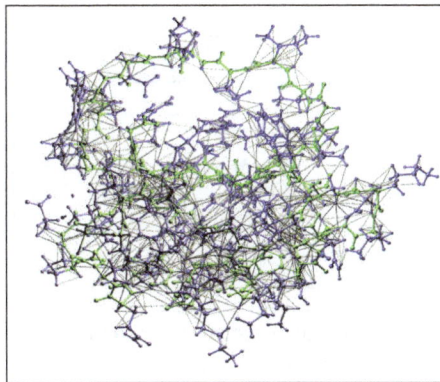

Illustration of the NOE based distance constraints used to determine the structure of the protein chymotrypsin inhibitor 2. The distance constraints are drawn as thin lines between the two atoms, whose distances are constrained. The backbone atoms and the bonds between them are green, the side chain atoms and the bonds between them are blue.

- Application of NMR Spectroscopy to Study Chemical and Properties of Proteins:

When an NMR signal of a given nucleus has been assigned in the NMR spectrum, the signal represents a specific reporter for the environment of the nucleus in the protein structure. The chemical shift is a very sensitive parameter, which might report even subtle conformational changes near the nucleus. The scalar coupling, represented by the coupling constant may report on conformational changes in dihedral angles. The binding of a ligand or the pH dependent change of a charge may result in conformational changes and a change in the NMR parameters for the nuclei in the environment of the chemical modification. Since NMR is specific to the atomic level NMR is by far superior to any other spectroscopic technique to register the changes, which occur as a result of the protein binding to another molecule.

Chemical Exchange by NMR

When a ligand binds to a protein, the equilibrium of the interaction is determined by the rate constants for the formation of the complex and rate constants for the dissociation of the complex. It is often of interest to characterize the properties of the protein-ligand complex, and knowing the on and off rates of the reaction is an important part of this characterization. NMR can measure reaction rates. This is illustrated in figure below.

In a sample where a ligand is in slow exchange with the protein and the signal of a given nucleus has a chemical sifts for the bound conformation and a chemical shift for the free conformation. NMR will record the sample as containing two different species the bound and the free form, bottom situation in figure. If this sample is heated and the rate constants increase the two signals get broader. This is because the nucleus is rapidly transferred from one magnetization condition to another, leading to line broadening. As it is seen in the middle panel the line broadening effect is so strong that the NMR signal essentially disappears. At even higher exchange rates the transfer of magnetization is faster than the difference in the chemical shift frequencies of the two different exchange sites. The magnetization will not precess with either frequency but will be observed as an average. The position of the average chemical shift depends on the fractions of the time the nuclear spin spends in each of the sites.

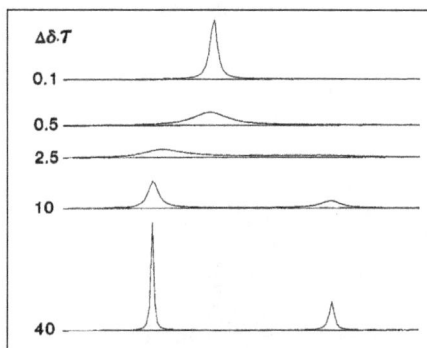

A model calculation of a two-site exchange system for the ratio between the chemical shift difference $\Delta\delta$ and the rate constant $1/\tau$ varying between 40 and 0.1.

The two extremes of fast and slow exchange, respectively, are both encountered in studies of protein ligand interactions when studied by NMR. In the fast exchange situation the ligand titration will behave as seen in figure.

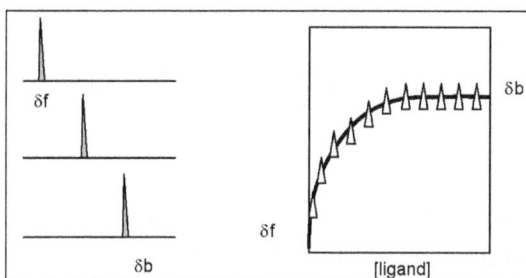

Schematic presentation of a "fast exchange" protein ligand titration by NMR. The top left: spectrum represents a start situation where no ligand has been added. The NMR signal is at δf. The middle line is the situation where enough ligand has been added to saturate half of the binding sites in the protein. The bottom line: ligand has been added to occupy all binding sites and the NMR signal is now at δb. right. Binding curve of a titration. The binding constant can be determined from the binding curve.

In the fast exchange titration the signal will shift with increasing concentration from the chemical shift position of the free protein, δf, to the chemical shift position δb. It is possible to determine the binding constant for the protein ligand interaction by fitting the observed binding curve to the theoretical expression for the binding. It is also possible from a line shape analysis to determine on- and off-rate constants.

Protein ligand interactions in slow exchange gives a completely different titration pattern.

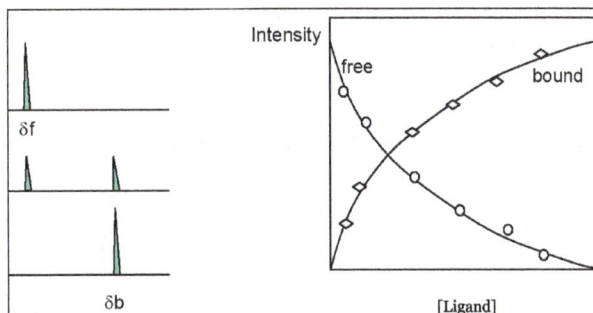

Schematic presentation of a "slow exchange" protein ligand titration. Left: change in spectrum with increasing ligand concentration. Right: change in signal intensities with ligand concentration.

Here the signal of the protein ligand complex will increase with the addition of ligand from no intensity to full intensity when all sites are occupied. At the same time intensity of the free protein signal will decrease. The sum of the intensity of the two peaks will always be constant. In principle the binding constant can be obtained from binding curves.

Application of NMR Spectroscopy for Studies of Chemical Reactions

For chemical processes, where chemicals are being produced, NMR spectroscopy is an ideal method for studying the time courses of the formation of a product and of the disappearance of the starting material. An NMR spectrum can be recorded in a fraction of a second, and it is therefore possible to study processes with half-life times from a second and up.

Enzymatic processes are often easily studied by NMR. One enormous advantage of NMR spectroscopy is its ability to study biochemical processes directly in living organisms.

Hydrogen Exchange in Proteins

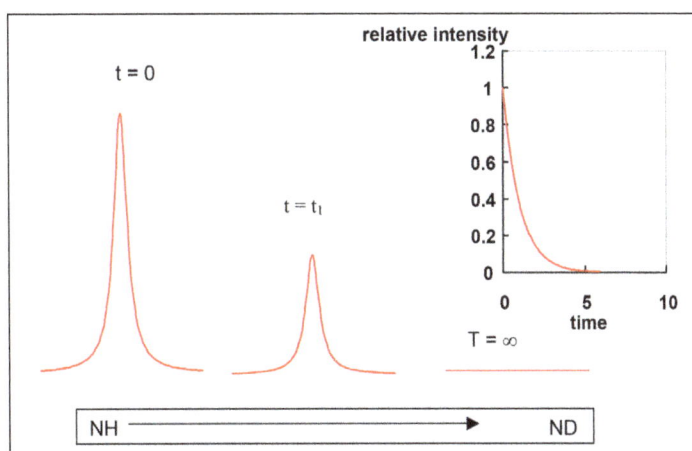

The decrease of the NH NMR signal with time in the process of the proton being exchanged with deuteron, which is not detectable by NMR at the protein frequency.

In aqueous solutions the hydrogen atom bound to nitrogen, oxygen and sulphur is labile and exchange with the hydrogen of water. In proteins this is the case for the peptide group, the amino group, the imidazol group, the indole group, the guanidine group, the hydroxy group and the thiol group. For proteins where these groups may engage in hydrogen bonds it is required, that the hydrogen bond is broken for the exchange to take place. The rate of the hydrogen exchange will depend of the frequency by which the hydrogen bond is opening.

Hydrogen exchange can be studied by NMR spectroscopy. By replacing water with deuterium oxide the exchange process will lead to the replacement of hydrogen with deuterium. 1H NMR can study the exchange process, since the hydrogen giving rise to a ^{1}H NMR signal is exchanged by deuterium, which is not observable by NMR at the ^{1}H NMR frequency.

The ^{1}H NMR signals of the peptide hydrogen atoms in a protein are normally very easily observed. These hydrogen atoms are engaged in hydrogen bond formation in the secondary structures of proteins. By recording the hydrogen exchange rates the stability of the individual hydrogen bonds can be measured. This can be done for every single peptide group of a protein and be used to describe the dynamic properties in the neighbourhood of the amide group, which is being monitored.

References

- Protein-Purification: labome.com, Retrieved 11 February, 2019

- Electrophoresis-of-proteins: intechopen.com, Retrieved 1 July, 2019

- Purification: ispybio.com, Retrieved 18 January, 2019

- Wang X, Hunter AK, Mozier NM (June 2009). "Host cell proteins in biologics development: Identification, quantitation and risk assessment". Biotechnology and Bioengineering. 103 (3): 446–58. doi:10.1002/bit.22304. PMID 19388135

- Amino-Acid-Sequencing, Determination-of-Primary-Structure, Nitrogen-Containing-Polymers-in-Nature, Amino-Acids, Peptides, Proteins, Organic-Chemistry: libretexts.org, Retrieved 8 August, 2019

- Spriestersbach A, Kubicek J, Schäfer F, Block H, Maertens B (2015). "Purification of His-Tagged Proteins". Methods in Enzymology. Elsevier. 559: 1–15. doi:10.1016/bs.mie.2014.11.00. ISBN 9780128002797. PMID 26096499

Chapter 3

Protein-Protein Interactions: Methods for Detection and Analysis

Protein–protein interactions (PPIs) are the physical contacts of high specificity which are established between two or more protein molecules. They are caused by biochemical events caused by electrostatic forces including the hydrophobic effect. The chapter closely examines these key concepts of protein-protein interactions to provide an extensive understanding of the subject.

Protein-Protein Interactions

Proteins control all biological systems in a cell, and while many proteins perform their functions independently, the vast majority of proteins interact with others for proper biological activity. Characterizing protein–protein interactions through methods such as co-immunoprecipitation (co-IP), pull-down assays, crosslinking, label transfer, and far–western blot analysis is critical to understand protein function and the biology of the cell.

Proteins are the workhorses that facilitate most biological processes in a cell, including gene expression, cell growth, proliferation, nutrient uptake, morphology, motility, intercellular communication and apoptosis. But cells respond to a myriad of stimuli, and therefore protein expression is a dynamic process; the proteins that are used to complete specific tasks may not always be expressed or activated. Additionally, all cells are not equal, and many proteins are expressed in a cell type–dependent manner. These basic characteristics of proteins suggest a complexity that can be difficult to investigate, especially when trying to understand protein function in the proper biological context.

Critical aspects required to understand the function of a protein include:

- Protein sequence and structure—used to discover motifs that predict protein function.

- Evolutionary history and conserved sequences—identifies key regulatory residues.

- Expression profile—reveals cell-type specificity and how expression is regulated.

- Post-translational modifications—phosphorylation, acylation, glycosylation and ubiquitination suggest localization, activation and/or function.

- Interactions with other proteins—function may be extrapolated by knowing the function of binding partners.

- Intracellular localization—may allude to the function of the protein.

Until the late 1990s, protein function analyses mainly focused on single proteins. However, because the majority of proteins interact with other proteins for proper function, they should be studied in the context of their interacting partners to fully understand their function. With the publication of the human genome and the development of the field of proteomics, understanding how proteins interact with each other and identifying biological networks has become vital to understanding how proteins function within the cell.

Types of Protein–protein Interactions

Protein interactions are fundamentally characterized as stable or transient, and both types of interactions can be either strong or weak. Stable interactions are those associated with proteins that are purified as multi-subunit complexes, and the subunits of these complexes can be identical or different. Hemoglobin and core RNA polymerase are examples of multi-subunit interactions that form stable complexes.

Transient interactions are expected to control the majority of cellular processes. As the name implies, transient interactions are temporary in nature and typically require a set of conditions that promote the interaction, such as phosphorylation, conformational changes or localization to discrete areas of the cell. Transient interactions can be strong or weak, and fast or slow. While in contact with their binding partners, transiently interacting proteins are involved in a wide range of cellular processes, including protein modification, transport, folding, signaling, apoptosis and cell cycling. The following example provides an illustration of protein interactions that regulate apoptotic and anti-apoptotic processes.

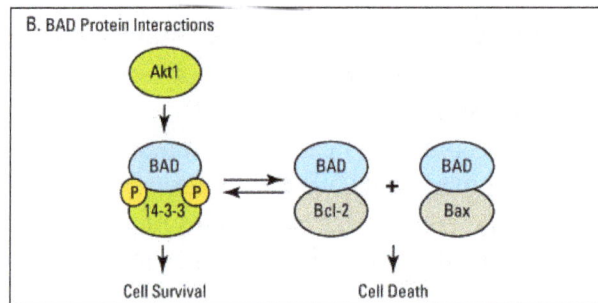

Heavy BAD protein–protein interaction. Panel A: Coomassie-stained SDS-PAGE gel of recombinant light and heavy BAD-GST-HA-6xHIS purified from HeLa IVT lysates (L), using glutathione resin (E1) and cobalt resin (E2) tandem affinity. The flow-through (FT) from each column is indicated. Panel B: Schematic of BAD phosphorylation and protein interactions during cell survival and cell death (i.e., apoptosis). Panel C: BAD protein sequence coverage showing identified Akt consensus phosphorylation sites (red box). Panel D: MS spectra of stable isotope-labeled BAD peptide HSSYPAGTEDDEGmGEEPSPFr.

Proteins bind to each other through a combination of hydrophobic bonding, van der Waals forces, and salt bridges at specific binding domains on each protein. These domains can be small binding clefts or large surfaces and can be just a few peptides long or span hundreds of amino acids. The strength of the binding is influenced by the size of the binding domain. One example of a common surface domain that facilitates stable protein–protein interactions is the leucine zipper, which consists of α-helices on each protein that bind to each other in a parallel fashion through the hydrophobic bonding of regularly-spaced leucine residues on each α-helix that project between the adjacent helical peptide chains. Because of the tight molecular packing, leucine zippers provide stable binding for multi-protein complexes, although all leucine zippers do not bind identically due to non-leucine amino acids in the α-helix that can reduce the molecular packing and therefore the strength of the interaction.

Two Src homology (SH) domains, SH2 and SH3, are examples of common transient binding domains that bind short peptide sequences and are commonly found in signaling proteins. The SH2 domain recognizes peptide sequences with phosphorylated tyrosine residues, which are often indicative of protein activation. SH2 domains play a key role in growth factor receptor signaling, during which ligand-mediated receptor phosphorylation at tyrosine residues recruits downstream effectors that recognize these residues via their SH2 domains. The SH3 domain usually recognizes proline-rich peptide sequences and is commonly used by kinases, phospholipases and GTPases to identify target proteins. Although both SH2 and SH3 domains generally bind to these motifs, specificity for distinct protein interactions is dictated by neighboring amino acid residues in the respective motif.

Biological Effects of Protein–protein Interactions

The result of two or more proteins that interact with a specific functional objective can be demonstrated in several different ways. The measurable effects of protein interactions have been outlined as follows:

- Alter the kinetic properties of enzymes, which may be the result of subtle changes in substrate binding or allosteric effects.

- Allow for substrate channeling by moving a substrate between domains or subunits, resulting ultimately in an intended end product.

- Create a new binding site, typically for small effector molecules.

- Inactivate or destroy a protein.

- Change the specificity of a protein for its substrate through the interaction with different binding partners, e.g., demonstrate a new function that neither protein can exhibit alone.

- Serve a regulatory role in either an upstream or a downstream event.

Common Methods to Analyze Protein–protein Interactions

Usually a combination of techniques is necessary to validate, characterize and confirm protein interactions. Previously unknown proteins may be discovered by their association with one or more proteins that are known. Protein interaction analysis may also uncover unique, unforeseen functional roles for well-known proteins. The discovery or verification of an interaction is the first step on the road to understanding where, how and under what conditions these proteins interact in vivo and the functional implications of these interactions.

While the various methods and approaches to studying protein–protein interactions are too numerous to describe here, the table below and the remainder of this topic focuses on common methods to analyze protein–protein interactions and the types of interactions that can be studies using each method. In summary, stable protein–protein interactions are easiest to isolate by physical methods like co-immunoprecipitation and pull-down assays because the protein complex does not disassemble over time. Weak or transient interactions can be identified using these methods by first covalently crosslinking the proteins to freeze the interaction during the co-IP or pull-down. Alternatively, crosslinking, along with label transfer and far–western blot analysis, can be performed independent of other methods to identify protein–protein interactions.

In Vitro Techniques

Co-immunoprecipitation

Co-Immunoprecipitation (Co-IP) is a classic technology widely used for protein-protein interaction identification and validation. Based on the specific immunological interaction between the bait protein and its antibody, co-IP has become an effective and reliable method in detecting the physiological interaction between proteins.

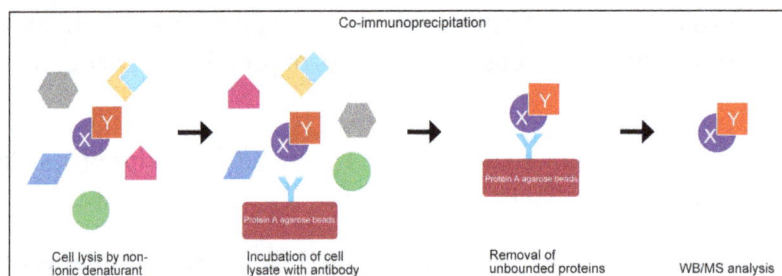

Co-immunoprecipitation

Cell lysis by non-ionic denaturant · Incubation of cell lysate with antibody · Removal of unbounded proteins · WB/MS analysis

Principles

As Co-Immunoprecipitation originates from immunoprecipitaiton techniques, it shares the same principle of antigen-antibody specificity/reaction as in any Immunoprecipitation technique. When a cell is lysed in non-denaturing condition, it retains the proteins in their native complex. Thus, according to the image the known protein in the complex X is referred as bait protein and the interacting protein Y is known as prey protein respectively. In a normal cellular environment, protein X will get attached to protein Y and the complex protein-protein Y is precipitated or isolated in presence of an antibody which can bind specifically to protein X. At the end, Co-IP thus gives information whether protein X and Y are interacting with each other or Y as a new interacting partner of X and also provide supports related to the information. The method starts with the lysis of cells or tissues which are known to have the protein or interacting partners of interest.Lysis buffers which contains non.ionic detergents are thus preferred as to retain the native ineractions status. The Bait protein is isolated from the lysed sample and immunecomplex containg prey proteins thus get captured with the bait protein. The immunecomplex containing a mixture of Bait protein, Bait-Prey protein and antibody is gets precipitated while using beads such as sepharose, agarose or magnet. A washing step next frees the resulting immunocomplex from irrelevant binding proteins, non-binding proteins. Further the precipitated partners are identified by methods like SDS-PAGE separation or mass spectrometry. In principal, Co-IP can be demonstrated as an extension on IP methods for isolating and purifying targets which identify secondary targets like interacting proteins in addition to primary antigen.

Applications

Immunoprecipitation method was originally invented as an alternate to affinity chromatography for small scale protein production. This method is a popular choice for identifying and separating protein with low abundance. Co-immunoprecipitation is a well known and popular method for studying protein-protein ineraction invitro. The main applications are:

- Prove interaction between two proteins.

- Identifies a new protein by using a known one.

Advantages

- In Co-IP method proteins which are interacting are generally Post-translationally modified and conformationally natural.

- Proteins interact in almost physiological condition.

- This method is highly specific in nature and at the same time quite simple to perform.

- The eluted protein complex is compatible with all applications which are used downstream such as western blot or mass spectrometry.

- Reagents which are used for Co-IP can be used several time thus making it cost effective.

Disadvantages

- Weaker signals from low affinity proteins are not detected.

- Not suitable for identification of protein interaction which takes place within short time period.

- Antibody selection is critical and target protein prediction needs to be correct.

- There could be a chance that two proteins are not directly interacting each other. There might ne presence of a third protein and the presence of such third protein in certain protein-protein interaction could weaken the procedure.

- Antibodies with high affinity or avidity are often difficult to isolate.

- The interacting protein identification is mostly based on the prediction of interaction between Bait and Prey antibody. This a wrong interpretation of Bait antibody will results no interacting partner while performing the method.

Protocol

- 15ml centrifuge tube,

- Microcenrifuge,

- Rotating shakers,

- Protein A/G beads,

- Equipments.

Workflow of Co-Immunoprecipitation

The steps of Co-immonuprecipitation goes through several steps such as cell lysis, pre cleaning of beads, binding of antibody to immune complex, washing, elution of target protein and detection of immune complex.

- Cell lysis: Cell lysis is the first step in the Co-immunoprecipitation method. The choice of buffer should be mild as they should not interfere with the protein-protein interaction, but at the same time required to be harsh so the buffer should be able to extract out the protein.

- Pre-cleaning: Pre-cleaning procedure is important for reducing background generated by immune complex formation. Without any valid reason, pre-cleaning step should not be omitted.

- Binding of immune complex: Binding includes complex formation between antigen, antibody and beads. Buffers which is used in immunecomplex formation and washing are the critical factor for binding process. The order, how the components are attaching to each other is a crucial factor for successful Co-IP. One way is to add the antibody to the beads and next add the lysate to the antibody-bead complex. Otherwise, antibody is first incubated with the cell lysate to form antibody-antigen complex and next beads are added to pull down the immune complex.

- Washing: Washing is an important step, as this will wash of the non-specifically binding protein retaining the specific immune partners. If background problem persists while detecting, a more stringent wash buffer will help to obtain purified complex.

- Elution of protein complex: For elution harsh environment such as loading buffer with SDS, or mild elution buffer like Glycine buffer is used.

- Detection: Interacting protein partners are detected by methods like Western blot or Mass spectrometry. Western blot method is based on the detection of desired protein with the help of detection antibody. As the amount of precipitated protein is not too high, a sensitive substrate is required for western blot technique.

On the other hand immunoprecipitation in a combination of mass sprectrometry or IP-MS is a direct method of identifying the target protein with the help of peptide sequence.

Controls in Co-immunoprecipitation: For a successful Co-immunoprecipitation experiment, incorporation of positive and negative controls is necessary. Total cell lysate can be used as positive control. Alternatively and protein which is not related to target protein can also use as negative control. Positive control confirms whether the assay was successful or not. Whereas, empty beads combined with cell lysate without the antibody can be used as negative control. When there is no bound IgG is present with the lysate-bead complex, it should not generate any bands in detection method like western blot. Alternatively antibody used in immune complex capturing can be added to the lysate without the beads, as this will indicate antibody cross reactivity. Another approach is to bind empty beads with lysate to check if there is any interference is created from the beads.

Detailed Procedure

The process starts with the transfection of cells with the protein expressing plasmid for the production of Bait protein or protein of interest. Transfection protocol is required to follow as per the instruction of transfection agents.

1. Generally after 48 hours the transfected cultured adherent cells are carefully washed two times with chilled PBS twice. Non adherent cells are washed with PBS and centrifuge at the rate of maximum of 800-1000 rpm for five minutes for pelleting the cells.

2. Next, ice cold RIPA Lysis buffer with protease inhibitor cocktail is added to the culture plate (1ml for 10^7 cells).

3. The cells are scrapped off from the culture plate to a5ml precooled Eppendorf tubes with a clean and ice cold scraper.

4. After this step, the cell suspension is incubated in low-speed rotating shaker for 15 mins at 4 °C to lyse the cells.

5. Next, the cell suspension is centrifuged in the eppendorf at 14,000 g 4 °C for 15min, the supernatant is transferred to new tubes immediately and placed on ice. The pellet is discarded. The supernatant can be stored as -80°C for a longer time period.

The critical parameter for successful Co-IP is to maintain the protein-protein interaction in their native physiological state. Thus all work with cell lysis buffer should be performed at low temperature, and in the presence of non-ionic detergents in lysis buffer. During a Co-IP assay, location of target protein should also be considered. For instance, target proteins which are located inside the organelle are required to release before the protein is interacting with the antibody. On the other hand, if the target protein is located on the membrane, their extraction will depend on the detergent type and concentration. Types of detergent in lysis buffer are also dependent of property of target protein.

Centrifugation force applied to make the cell supernatant may require to adjust based on cell type.

6. In the meantime, agarose beads are prepared for the next step. Generally, the agarose A/G beads are resin beads which are used for purification of antibody these beads are consisting of recombinant fusion protein A/G which has been immobilized on 6% agarose and are suitable for purification of IgG immunoglobulins. Usually the protein A/G-agarose beads are washed twice with PBS and a 50% protein slurry of A/G agarose working solution (in PBS) is made. It is advisable to cut the front of the tip, so thus beads will not be disrupted.

7. Next, 50% protein slurry of A/G agarose is added with ratio of 100µl for a 1ml sample solution (supernatant).

8. After this the eppendorf tube containing supernatant with A/G agarose bead is set to incubate in horizontal shaker for 10min at 4 °C. This step is done mainly to eliminate non-specific binding proteins.

9. Again, the tube is Centrifuged 14,000g at 4 °C for 15min.

10. At this stage, the supernatant is transferred to new centrifuge tubes and discard protein A/G-agraose beads.

Beads at this stage can be washed with lysis buffer to increase the amount of protein complex.

11. The protein is next quantified by BCA assay or other protein quantification.

A SDS gel can be run at this stage. Coomasie staining of the gel will show whether serum Ig has been removed properly from the bead-antigen complex. Presence of bands at the molecular weight of 25 and 50Kd indicates presence of antibody heavy and light chain which may hinder the immunoprecipitaiton in the next steps.

12. The protein is further diluted to 1µg/µl with PBS to decrease the concentration of detergents.

13. The diluted primary antibody is added at an appropriate amount to 500µl total volume of the solution of supernatant-bead complex.

14. Next, the antigen-antibody complex in centrifuge tube is incubated for overnight at 4 °C in a rotating shaker.

For downstream enzyme assay after immunoprecipitaiton, a shorter incubation time is recommended.

15. Centrifuge the tube at 14,000g for 5s, keep the pellet and wash with pre-chilled washing buffer for 3 times (800µl each).

16. The supernatant is collected and analysis is carried out via SDS-PAGE, western-blot, or mass spectra analysis. Alternatvely the supernatant can also be stored at 4 °C for future application.

This protocol gives lower yield, but avoids the problem of co-elution of antibodies. If higher yield is expected, the antibody can be mixed with protein sample prior to addition of Protein A/G-agarose beads. The antibodies can co-elude with target protein and interference can be observed in western blot detection.

Preparation of Reagents and Buffers

1. Phosphate Buffered saline, 1X PBS.

2. RIPA lysis buffer is made of Tris-HCl: 50 mM, pH 7.4, Nonidet P-40 (NP-40): 1%, Deoxycholate Na: 25%, NaCl: 120 mM, EDTA: 1 mM, PMSF: 1 mM, Leupeptin 1 µg/ml, Aprotinin 1 µg/ml, Pepstatin1 µg/ml, Na3VO4: 1 mM, and NaF: 1 mM. (PMSF, leupeptine, Aprtonin, pepstatin should be added freshly to the lysus buffer Amount the NaCl should no exceed 1M.

3. Washing Buffer made of tris buffer saline with tween 20.

4. Protein A/G-agarose beads.

5. Specific antibody (MAb or PAb).

6. Elution buffer: SDS-PAGE sample loading buffer or Glycine buffer.

Optimization of Co-Immunoprecipitaiton

Although the principal of Co-IP is easy to perform, however, there are few circumstances which may interfere with the protein-protein interaction in physiologically relevant conditions.

Binding: As in Co-IP cells go through mechanical and chemical procedures, incubations and washing steps, it is very necessary that the protein-protein interactions maintain stability. Thus in situations where the protein-protein interaction is transient or affinity in low, Co-IP might not perform able to detect the interaction partners. Thus an important step for retaining the interaction is cell lysis and washing. While choosing the buffer it is important to choose low ionic strength buffer with detergents which are non-ionic in nature like NP-40 so the target protein complex remain unaffected. In addition, gentle handling is also required during centrifugation and washing steps to prevent loss of protein complex. During the assay period, it is essential to keep the lysate or supernatant on ice and addition of protease and phosphatase inhibitors are highly recommended to as there are chance of proteolysis or dephosphorylatiion of proteins exists. Repeating freezw thaw cycle can damage the protein-protein interaction, thus working with fresh lysates are the best option for desired result production.

Nonspecific binding: As in the cells there are lot of proteins it is inevitable that nonspecific binding will takes place. In addition to this, during the lysis steps cellular compartmentalization of a cell get destroyed, there is a high chance that proteins like actin which is highly abundant might bind and interfere with target protein. Washing is thus most important criteria to get rid from the nonspecific interactions. However, other options such as using higher ionic strength lysis buffer or lower amount of primary anibody or a precleaning step of cell lysate may reduce the risk of nonspecific binding.

Selection of antibody: As antibody is the important part of the immunoprecipitation assay, is it always advisable to use antibody of different origin for the capturing and detecting methods, such as anti-rabbit antibody for the pulling down of protein complex and anti-mouse antibodies for downstream detection like western blots. This will help to reduce the chance of interference generated from heavy and light chain of the antibody which is used to isolate the protein complex. Apart from use of antibodies from different species, polyclonal antibodies are also preferred for a higher binding of antibody to protein complex. In comparison to monoclonal antibodies, polyclonal antibodies have the capacity to get attached with the multiple epitopes expressed in target proteins. This will lowered the chance of protein complex being washed away during the washing steps. Whereas during detection, monoclonal antibody is preferred for specific detection.

Contamination of antibody: Antibody contamination is a common problem in Co-immunoprecipitation. Downstream applications such as Western Blot may generate additional bands in the gel. The light and heavy chain antibodies which might have co-eluted with the protein complex and interfere with the identification process. One way to get rid of the problem is covalent binding of antibody with the beads. This method also allows the reuse of beads coated with antibody. Another strategy used to reduce cross contamination of antibody is use of biotin-streptavidin system, where biotinylated antibody is used for target protein whereas streptavidin is used to coat the beads.

Choice of beads: There are few types of beads available for the use in immunoprecipitation methods. Among them protein A and protein G conjugated beads are widely used. Both protein A and G has high affinity towards antibody, but depending on the cell type and target organisms, they will show difference in antibody binding. In general protein A beads are working better with rabbit antibodies, while protein G for mouse antibodies. Apart from protein A/G beads, agarose and magnetic beads are commonly used in Co-IP. Agarose beads have a porous sponge like structure and they shows high affinity towards antibody. The use of agarose beads is also cost effective. However, as antibodies get highly trapped in this porous structure and become difficult to wash out. Thus while using agarose beads extensive cell lysate cleaning and washing is required for background reduction. Additionally agarose beads are also required longer incubation period as porous structures results in lower diffusion rate. On the other hand, magnetic beads are small with smoother outside. As a result they have low binding capacity and required a large volume of beads. But they do not require any centrifugation and less washing steps. Magnetic beads showed increased diffusion rate and results in lower incubation time with antibody. Altogether they are a good option for used in Co-immunoprecipitation.

Protein Microarrays

Protein microarrays, an emerging class of proteomic technologies, are fast becoming critical tools in biochemistry and molecular biology. Two classes of protein microarrays are currently available: Analytical and functional protein microarrays. Analytical protein microarrays, mostly antibody microarrays, have become one of the most powerful multiplexed detection technologies. Functional protein microarrays are being increasingly applied to many areas of biological discovery, including studies of protein interaction, biochemical activity, and immune responses. Great progress has been achieved in both classes of protein microarrays in terms of sensitivity, specificity, and expanded application.

Protein microarrays, also known as protein chips, are miniaturized and parallel assay systems that contain small amounts of purified proteins in a high-density format. They allow simultaneous determination of a great variety of analytes from small amounts of samples within a single experiment. Protein microarrays are typically prepared by immobilizing proteins onto a microscope slide using a standard contact spotter or noncontact microarrayer. A variety of slide surfaces can be used. Popular types include aldehyde-and epoxy-derivatized glass surfaces for random attachment through amines, nitrocellulose, or gel-coated slides and nickel-coated slides for affinity attachment of His6-tagged proteins. The last type was reported to provide 10-fold better signals than those obtained with other random attachment methods. After proteins are immobilized on the slides, they can be probed for a variety of functions/activities. Finally, the resulting signals are usually measured by detecting fluorescent or radio-isotope labels. The typical image of protein microarrays is shown as figure below.

A typical protein microarray image.

A yeast protein microarray is probed with anti-GST antibodies followed by detection with Cy5-conjugated secondary antibodies. An enlarged image of one of the 48 blocks is depicted below the protein chip.

Analytical protein arrays can be used to monitor protein expression levels or for bio-marker identification, clinical diagnosis, or environmental/food safety analysis. Functional protein microarrays have many uses: (i) To probe for various types of protein activities, including protein-protein, protein-lipid, protein-DNA, protein-drug, and protein-pep-tide interactions; (ii) To identify enzyme substrates; and (iii) To profile immune responses, among many others. Applications of both the analytical and functional protein microarrays are depicted in figure. In the following sections, we will provide examples of various applications of both types of microarray, with an emphasis on functional protein microarrays. Given the large volume of papers related to protein microarray technology, we regret that we are unable to cite all the published work in the field.

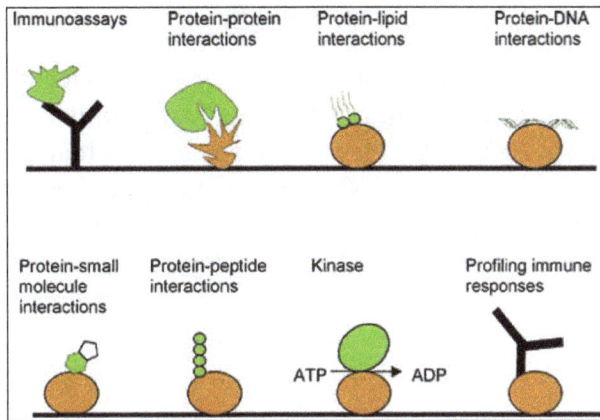

Applications of protein microarrays.

Antibody arrays can be used for clinical diagnosis or environmental/food safety analysis. Functional protein arrays are mainly used to study various types of protein activities, including protein-protein, protein-lipid, protein-DNA, protein-drug, and protein-peptide interactions, to identify enzyme substrates and to profile immune responses.

Analytical Microarrays

Perhaps the most representative class of analytical micro-arrays is the antibody microarray, in which antibodies are arrayed on glass surfaces at high density. The biggest challenge associated with antibody microarrays is that of producing antibodies that are able to identify the proteins of interest with high specificity and affinity in a high-throughput fashion. Because the traditional method for generating monoclonal antibodies is time-consuming and laborious, researchers have recently sought alternative approaches. For example, phage antibody-display, ribosome display, systematic evolution of ligands by exponential enrichment (SELEX), messenger RNA (mRNA) display, and affibody display have been developed to expedite the production of antibodies with high specificity. All of these methods involve the construction of large repertoires of viable regions with potential binding activity, which can be selected by multiple rounds of affinity purification. The binding affinity of the resulting candidate clones can be further improved using maturation strategies. However, the ideal selection system is yet to be fully developed: One that is not only fast, robust, sensitive, and of low cost, but also automated and minimized.

Despite the challenge involved in obtaining specific antibodies, many studies using antibody microarrays have recently been reported. In a pioneer work by Haab and colleagues, the first high-density antibody microarrays were used to test whether a linear relationship could be detected between an antibody and antigen pair in an array format. They investigated the ability of 115 well-characterized antibody-antigen pairs to react in high-density microarrays on modified glass slides: 30% of the pairs showed the expected linear relationships, indicating that a fraction of the antibodies were suitable for quantitative analysis. Sreekumar and coworkers created antibody arrays with 146 distinct antibodies against proteins involved in the stress response, cell cycle progression, and apoptosis and used these arrays to monitor the alterations in protein quantity in LoVo colon carcinoma cells. The reference standards and samples were labeled separately using either Cy™5 or Cy3 dyes, and the fluorescent signals of the bound proteins were detected with a confocal microarray scanner. These investigators were able to obtain differential expression profiles, with radiation-induced up-regulation of apoptotic regulators, such as p53, DNA fragmentation factors, and tumor necrosis factor-related ligand.

In order to increase affinity and specificity, analytical micro-arrays usually employ a signal amplification system and sandwich assay format, in which the first antibody is spotted on the array and then a captured antigen on the chip is detected with a second antibody that recognizes a different part of the antigen. A highly sensitive antibody microarray system combining both methods has been shown to be capable of simultaneously detecting 75 cytokines with high specificity, femtomolar sensitivity, a 3-log quantitative range, and economy of sample consumption. Although the sandwich format dramatically increases the specificity of the antigen detection, it requires at least two high-quality antibodies for each antigen that is to be detected.

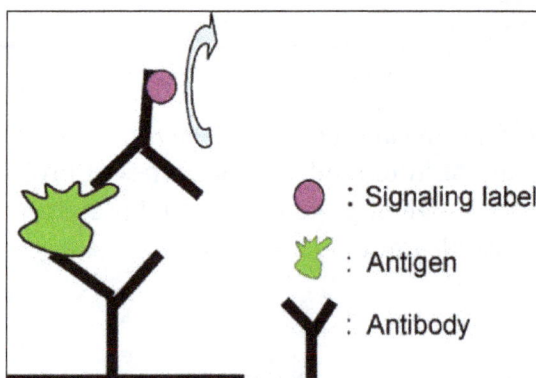

A sandwich assay format.

A multivalent antigen is first caught by a capture antibody immobilized on the surface and then detected by a detection antibody. The label is usually tagged on the detection antibody and can be further amplified.

Functional Protein Microarrays

Functional protein microarrays have recently been applied to many aspects of discovery-based biology, including protein-protein, protein-lipid, protein-DNA, protein-drug, and protein-peptide interactions. Although we have attempted to describe all the major applications of functional protein microarrays, it is impossible to cover all the instances in which they have been used. Therefore, we have chosen to focus most of our examples on yeast proteome microarrays.

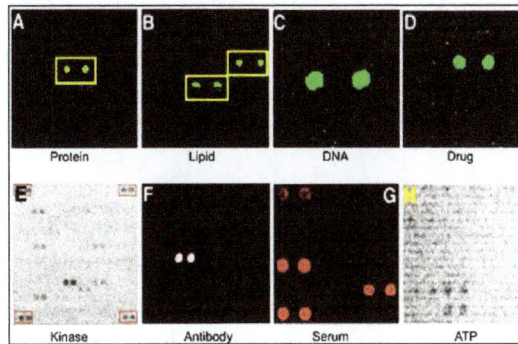

Examples of different assays on functional protein chips.

Different types of biochemical assays were carried out on chips, including assays of (A) protein-protein, (B) protein-lipid, (C) protein-DNA, (D) protein-drug, (H) protein-small molecule, and (F) protein-antibody interactions. The chips can also be used to monitor immune responses in patients (G) and posttranslational modifications of proteins, such as phosphorylation (E). These assays achieved high signal-to-noise ratios and were very informative for elucidating the function of previously uncharacterized genes.

Phage Display

Phage display is a laboratory platform that allows scientists to study protein interactions on a large-scale and select proteins with the highest affinity for specific targets.

The photograph shows a phage as seen under a microscope.

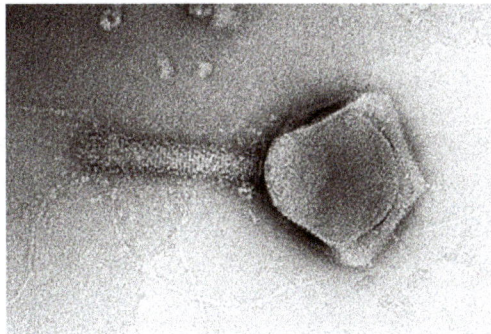

Importance

The key advantage of phage display is that it provides a means to identify target-binding proteins from a library of millions of different proteins without the need to screen each molecule individually. This makes it possible to screen billions of proteins each week. By linking a selected protein with its encoding gene, phage display also provides a means to easily identify coding sequences of binding proteins. These can be stored, amplified or processed in other ways. Phage display is a pivotal tool for early basic scientific research and for the development of new drugs and vaccines. The technology has proved particularly important to the production of safer and more effective monoclonal antibody drugs. Phage display libraries consisting entirely of human antibody sequences, for example, have made it possible to produce fully human antibodies. Four fully human therapeutic antibodies currently approved as treatments in the US and the UK were developed using phage display and many more are in the pipeline. Phage display is also used to develop vaccines for conditions such as prostate cancer and HIV. It is also an important tool for the generation of diagnostic tests for monitoring disease progression and evaluating treatment efficacy.

Principle

phage display is used for the high-throughput screening of protein interactions. In the case of M13 filamentous phage display, the DNA encoding the protein or peptide of interest is ligated into the pIII or pVIII gene, encoding either the minor or major coat protein, respectively. Multiple cloning sites are sometimes used to ensure that the fragments are inserted in all three possible reading frames so that the cDNA fragment is translated in the proper frame. The phage gene and insert DNA hybrid is then inserted (a process known as "transduction") into Escherichia coli (E. coli) bacterial cells such as TG1, SS320, ER2738, or XL1-Blue E. coli. If a "phagemid" vector is used (a simplified display construct vector) phage particles will not be released from the E. coli cells until they are infected with helper phage, which enables packaging of the phage DNA and assembly of the mature virions with the relevant protein fragment as part of their outer coat on either the minor (pIII) or major (pVIII) coat protein. By immobilizing a relevant DNA or protein target(s) to the surface of a microtiter plate well, a phage that displays a protein that binds to one of those targets on its surface will remain while others are removed by washing. Those that remain can be eluted, used to produce more phage (by bacterial infection with helper phage) and to produce a phage mixture that is enriched with relevant (i.e. binding) phage. The repeated cycling of these steps is referred to as 'panning', in reference to the enrichment of a sample of gold by removing undesirable materials. Phage eluted in the final step can be used to infect a suitable bacterial host, from which the phagemids can be collected and the relevant DNA sequence excised and sequenced to identify the relevant, interacting proteins or protein fragments.

The use of a helper phage can be eliminated by using 'bacterial packaging cell line' technology.

Elution can be done combining low-pH elution buffer with sonication, which, in addition to loosening the peptide-target interaction, also serves to detach the target molecule from the immobilization surface. This ultrasound-based method enables single-step selection of a high-affinity peptide.

Applications

Applications of phage display technology include determination of interaction partners of a protein (which would be used as the immobilised phage "bait" with a DNA library consisting of all coding sequences of a cell, tissue or organism) so that the function or the mechanism of the function of that protein may be determined. Phage display is also a widely used method for in vitro protein evolution (also called protein engineering). As such, phage display is a useful tool in drug discovery. It is used for finding new ligands (enzyme inhibitors, receptor agonists and antagonists) to target proteins. The technique is also used to determine tumour antigens (for use in diagnosis and therapeutic targeting) and in searching for protein-DNA interactions using specially-constructed DNA libraries with randomised segments. Recently, phage display has also been used in the context of cancer treatments - such as the adoptive cell transfer approach. In these cases, phage display is used to create and select synthetic antibodies that target tumour surface proteins. These are made into synthetic receptors for T-Cells collected from the patient that are used to combat the disease.

Competing methods for in vitro protein evolution include yeast display, bacterial display, ribosome display, and mRNA display.

Antibody Maturation in Vitro

The invention of antibody phage display revolutionised antibody drug discovery. Initial work was done by laboratories at the MRC Laboratory of Molecular Biology, the Scripps Research Institute and the German Cancer Research Centre. In 1991, The Scripps group reported the first display and selection of human antibodies on phage. This initial study described the rapid isolation of human antibody Fab that bound tetanus toxin and the method was then extended to rapidly clone human anti-HIV-1 antibodies for vaccine design and therapy.

Phage display of antibody libraries has become a powerful method for both studying the immune response as well as a method to rapidly select and evolve human antibodies for therapy. Antibody phage display was later used by Carlos F. Barbas at The Scripps Research Institute to create synthetic human antibody libraries, a principle first patented in 1990 by Breitling and coworkers (Patent CA 2035384), thereby allowing human antibodies to be created in vitro from synthetic diversity elements.

Antibody libraries displaying millions of different antibodies on phage are often used in the pharmaceutical industry to isolate highly specific therapeutic antibody leads, for development into antibody drugs primarily as anti-cancer or anti-inflammatory therapeutics. One of the most successful was adalimumab, discovered by Cambridge Antibody Technology as D2E7 and developed and marketed by Abbott Laboratories. Adalimumab, an antibody to TNF alpha, was the world's first fully human antibody, which achieved annual sales exceeding $1bn.

Protocol

Below is the sequence of events that are followed in phage display screening to identify polypeptides that bind with high affinity to desired target protein or DNA sequence:

1. Target proteins or DNA sequences are immobilized to the wells of a microtiter plate.

2. Many genetic sequences are expressed in a bacteriophage library in the form of fusions with the bacteriophage coat protein, so that they are displayed on the surface of the viral particle. The protein displayed corresponds to the genetic sequence within the phage.

3. This phage-display library is added to the dish and after allowing the phage time to bind, the dish is washed.

4. Phage-displaying proteins that interact with the target molecules remain attached to the dish, while all others are washed away.

5. Attached phage may be eluted and used to create more phage by infection of suitable bacterial hosts. The new phage constitutes an enriched mixture, containing considerably less irrelevant phage (i.e. non-binding) than were present in the initial mixture.

6. Steps 3 to 5 are optionally repeated one or more times, further enriching the phage library in binding proteins.

7. Following further bacterial-based amplification, the DNA within in the interacting phage is sequenced to identify the interacting proteins or protein fragments.

Selection of the Coat Protein

pIII

pIII is the protein that determines the infectivity of the virion. pIII is composed of three domains (N1, N2 and CT) connected by glycine-rich linkers. The N2 domain binds to the F pilus during virion infection freeing the N1 domain which then interacts with a TolA protein on the surface of the bacterium. Insertions within this protein are usually added in position 249 (within a linker region between CT and N2), position 198 (within the N2 domain) and at the N-terminus (inserted between the N-terminal secretion sequence and the N-terminus of pIII). However, when using the BamHI site located at position 198 one must be careful of the unpaired Cysteine residue (C201) that could cause problems during phage display if one is using a non-truncated version of pIII.

An advantage of using pIII rather than pVIII is that pIII allows for monovalent display when using a phagemid (Ff-phage derived plasmid) combined with a helper phage. Moreover, pIII allows for the insertion of larger protein sequences (>100 amino acids) and is more tolerant to it than pVIII. However, using pIII as the fusion partner can lead to a decrease in phage infectivity leading to problems such as selection bias caused by difference in phage growth rate or even worse, the phage's inability to infect its host. Loss of phage infectivity can be avoided by using a phagemid plasmid and a helper phage so that the resultant phage contains both wild type and fusion pIII. cDNA has also been analyzed using pIII via a two complementary leucine zippers system, Direct Interaction Rescue or by adding an 8-10 amino acid linker between the cDNA and pIII at the C-terminus.

pVIII

pVIII is the main coat protein of Ff phages. Peptides are usually fused to the N-terminus of pVIII. Usually peptides that can be fused to pVIII are 6-8 amino acids long. The size restriction seems to have less to do with structural impediment caused by the added section and more to do with the size exclusion caused by pIV during coat protein export. Since there are around 2700 copies of the protein on a typical phages, it is more likely that the protein of interest will be expressed polyvalently even if a phagemid is used. This makes the use of this protein unfavorable for the discovery of high affinity binding partners.

To overcome the size problem of pVIII, artificial coat proteins have been designed. An example is Weiss and Sidhu's inverted artificial coat protein (ACP) which allows the display of large proteins at the C-terminus. The ACP's could display a protein of 20kDa, however, only at low levels (mostly only monovalently).

pVI

pVI has been widely used for the display of cDNA libraries. The display of cDNA libraries via phage display is an attractive alternative to the yeast-2-hybrid method for the discovery of interacting proteins and peptides due to its high throughput capability. pVI has been used preferentially to pVIII and pIII for the expression of cDNA libraries because one can add the protein of interest to the C-terminus of pVI without greatly affecting pVI's role in phage assembly. This means that the stop codon in the cDNA is no longer an issue. However, phage display of cDNA is always limited by the inability of most prokaryotes in producing post-translational modifications present in eukaryotic cells or by the misfolding of multi-domain proteins.

While pVI has been useful for the analysis of cDNA libraries, pIII and pVIII remain the most utilized coat proteins for phage display.

pVII and pIX

In an experiment in 1995, display of Glutathione S-transferase was attempted on both pVII and pIX and failed. However, phage display of this protein was completed successfully after the addition of a periplasmic signal sequence (pelB or ompA) on the N-terminus. In a recent study, it has been shown that AviTag, FLAG and His could be displayed on pVII without the need of a signal sequence. Then the expression of single chain Fv's (scFv), and single chain T cell receptors (scTCR) were expressed both with and without the signal sequence.

PelB (an amino acid signal sequence that targets the protein to the periplasm where a signal peptidase then cleaves off PelB) improved the phage display level when compared to pVII and pIX fusions without the signal sequence. However, this led to the incorporation of more helper phage genomes rather than phagemid genomes. In all cases, phage display levels were lower than using pIII fusion. However, lower display might be more favorable for the selection of binders due to lower display being closer to true monovalent display. In five out of six occasions, pVII and pIX fusions without pelB was more efficient than pIII fusions in affinity selection assays. The paper even goes on to state that pVII and pIX display platforms may outperform pIII in the long run.

The use of pVII and pIX instead of pIII might also be an advantage because virion rescue may be undertaken without breaking the virion-antigen bond if the pIII used is wild type. Instead, one could cleave in a section between the bead and the antigen to elute. Since the pIII is intact it does not matter whether the antigen remains bound to the phage.

T7 phages

The issue of using Ff phages for phage display is that they require the protein of interest to be translocated across the bacterial inner membrane before they are assembled into the phage. Some proteins cannot undergo this process and therefore cannot be displayed on the surface of Ff phages. In these cases, T7 phage display is used instead. In T7 phage display, the protein to be displayed is attached to the C-terminus of the gene 10 capsid protein of T7.

The disadvantage of using T7 is that the size of the protein that can be expressed on the surface is limited to shorter peptides because large changes to the T7 genome cannot be accommodated like it is in M13 where the phage just makes its coat longer to fit the larger genome within it. However, it can be useful for the production of a large protein library for scFV selection where the scFV is expressed on an M13 phage and the antigens are expressed on the surface of the T7 phage.

Pull-Down Assays

The pull-down assay is an in vitro method used to determine a physical interaction between two or more proteins. Pull-down assays are useful for both confirming the existence of a protein–protein interaction predicted by other research techniques (e.g., co-immunoprecipitation) and as an initial screening assay for identifying previously unknown protein–protein interactions.

Pull-down assays are a form of affinity purification and are similar to immunoprecipitation, except that a "bait" protein is used instead of an antibody. Affinity chromatography (i.e., affinity purification) methodologies greatly enhance the speed and efficiency of protein purification and simultaneously provide the technology platform to perform a pull-down, or co-purification, of potential binding partners. In a pull-down assay, a bait protein is tagged and captured on an immobilized affinity ligand specific for the tag, thereby generating a "secondary affinity support'" for purifying other proteins that interact with the bait protein. The secondary affinity support of immobilized bait is then incubated with a protein source that contains putative "prey" proteins, such as a cell lysate. The source of prey protein at this step depends on whether the researcher is confirming a previously suspected protein–protein interaction or identifying an unknown interaction. The method of protein elution depends on the affinity ligand and ranges from using competitive analytes to low pH or reducing buffers.

Besides investigating the interaction of two or more proteins, pull-down assays are a powerful tool to detect the activation status of specific proteins. For example, proteins that are activated in response to tyrosine phosphorylation can be pulled down using an immobilized SH2 domain that targets the phosphorylated tyrosine on a given protein. Additionally, GTPases, which act as molecular switches that regulate cell signaling by cycling between a GTP-bound (active) and GDP-bound (inactive) state, can be pulled down using an immobilized GTPase-binding domain of downstream proteins that are recruited to GTP-bound, activated GTPases. In both types of pull-down assays, because the specificity of the interaction is dependent on the sequence of the binding domain, these approaches are highly specific in detecting the activation of distinct proteins.

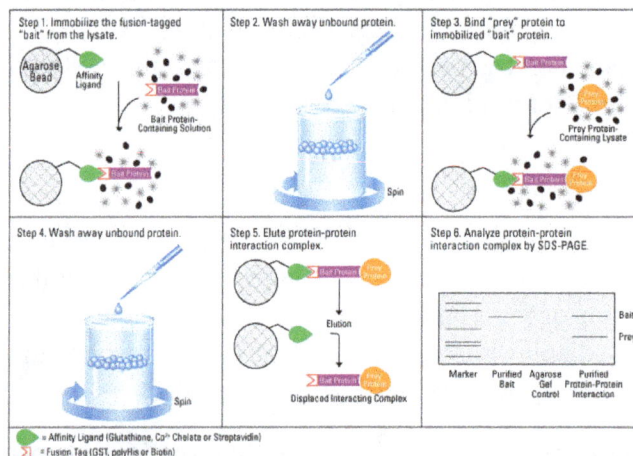

General schematic of a pull-down assay. A pull-down assay is a small-scale affinity purification technique similar to immunoprecipitation, except that the antibody is replaced by some other affinity system. In this case, the affinity system consists of a glutathione S-transferase (GST)–, polyHis- or streptavidin-tagged protein or binding domain that is captured by glutathione-, metal chelate (cobalt or nickel) – or biotin-coated agarose beads, respectively. The immobilized fusion-tagged protein acts as the "bait" to capture a putative binding partner (i.e., the "prey"). In a typical pull-down assay, the immobilized bait protein is incubated with a cell lysate, and after the prescribed washing steps, the complexes are selectively eluted using competitive analytes or low pH or reducing buffers for in-gel or western blot analysis.

The pull-down Assay as a Confirmatory Tool

The confirmation of previously suspected interactions typically utilizes a prey protein source that has been expressed in an artificial protein expression system. This allows the researcher to work with a larger quantity of the protein than is typically available under endogenous expression conditions and eliminates confusing results that could arise from interaction of the bait with other interacting proteins present in the endogenous system that are not under study. Protein expression system lysates (i.e., E. coli or baculovirus-infected insect cells), in vitro transcription/translation reactions, and previously purified proteins are appropriate prey protein sources for confirmatory studies.

The Pull-down Assay as a Discovery Tool

The discovery of unknown interactions contrasts with confirmatory studies because the research interest lies in discovering new proteins in the endogenous environment that interact with a given bait protein. The endogenous environment can entail a plethora of possible protein sources but is generally characterized as a complex protein mixture considered to be the native environment of the bait protein. Any cellular lysate in which the bait is normally expressed, or complex biological fluid (i.e., blood, intestinal secretions, etc.) where the bait would be functional, is an appropriate prey protein source for discovery studies.

Critical Components of Pull-down Assays

Bait Protein Criteria

Bait proteins for pull-down assays can be generated either by linking an affinity tag to proteins purified by traditional purification methods or by expressing recombinant fusion-tagged proteins. Researchers who have access to commercially available purified protein or frozen aliquots of purified protein from an earlier study can design a pull-down assay without the need for cloning the gene encoding the protein of interest. The purified protein can be tagged with a protein-reactive tag commonly used for such labeling applications. Alternatively, if a cloned gene is available, molecular biology methods can be employed to subclone the gene to an appropriate vector with a fusion tag. Recombinant clones can be overexpressed and easily purified, resulting in an abundance of bait protein for use in pull-down assays. In the representative example below, glutathione-S-transferase (GST) is utilized to perform a pull-down experiment.

Common Fusion Tags and their Affinity Binding Ligands

Fusion tag	Affinity ligand
Glutathione S-transferase (GST)	Glutathione
Poly-histidine (polyHis or 6xHis)	Nickel or cobalt chelate complexes
Biotin	Streptavidin

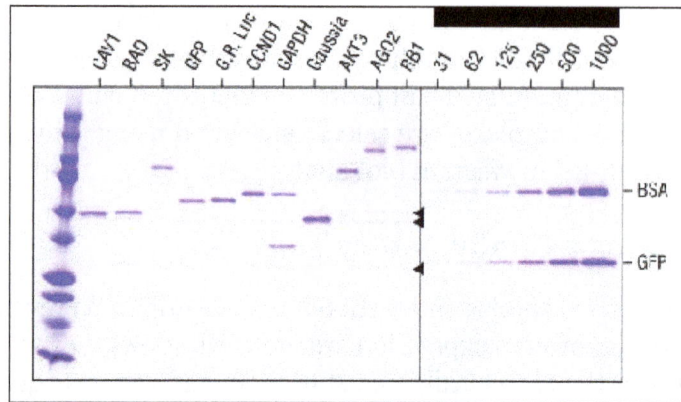

Purification of N-terminal GST fusion proteins with immobilized glutathione.

Binding Parameters: Stable vs. Transient Interactions

The discovery and confirmation of protein–protein interactions using the pull-down technique depend heavily on the nature of the interaction under study. Interactions can be stable or transient, and this characteristic determines the conditions for optimizing binding between the bait and prey proteins. Transient interactions are usually associated with transport or enzymatic mechanisms. The ribosome illustrates both examples because the structure consists of many stable protein–protein interactions, but the enzymatic mechanism that translates mRNA to nascent protein requires transient interactions.

Stable protein–protein interactions are easiest to isolate by physical methods like pull-down assays because the protein complex does not disassemble over time. Strong, stable protein complexes can be washed extensively with high ionic strength buffers to eliminate any false positive results due to nonspecific interactions. If the complex interaction has a higher dissociation constant and a weaker interaction, the interaction strength and thus the protein complex recovery can be improved by optimizing the assay conditions related to pH, salt species and salt concentration. Problems of nonspecific interactions can be minimized with the careful design of appropriate control experiments. This example provides an example of an experiment designed to pull-down the active form of Cdc42 using a specialized kit.

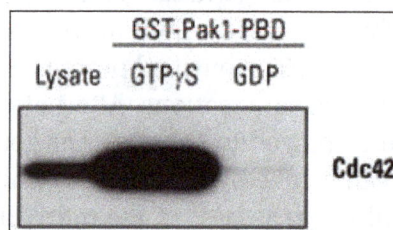

Detection of active Cdc42. The Active Cdc42 Pull-Down and Detection Kit was used to detect active protein present in NIH 3T3 cell lysate treated with GTPγS (activator) or GDP (inactivator). Active Cdc42 was enriched by pull-down assay. Half of each eluate (25 µL) and 40 µg of total lysate were analyzed by western blot using the anti-Cdc42 antibody supplied in the kit. Only active Cdc42 was detected, as indicated by the lower signal intensity in the GDP-treated sample.

Transient interactions are defined by their temporal interaction with other proteins and are the most challenging protein-protein interactions to isolate. These interactions are more difficult to identify using physical methods like pull-down assays because the complex may dissociate during the assay. Since transient interactions occur during transport or as part of enzymatic processes, they often require cofactors and energy via nucleotide triphosphates hydrolysis. Incorporating cofactors and nonhydrolyzable nucleoside triphosphate (NTP) analogs during assay optimization can serve to 'trap' interacting proteins in different stages of a functional complex that is dependent on the cofactor or NTP.

Weak or transient protein–protein interactions can be strengthened by covalently crosslinking the interacting proteins prior to pull-down. While this strategy is more advanced than performing the pull-down assay without crosslinking, freezing protein interactions by crosslinking may make or break the success of a pull-down assay.

Elution of the Bait–prey Complex

The identification of bait–prey interactions requires that the complex is removed from the affinity support and analyzed by standard protein detection methods. The entire complex can be eluted from the affinity support using sodium dodecyl sulfate–polyacrylamide gel electrophoresis (SDS-PAGE) loading buffer or a competitive analyte specific for the tag on the bait protein. SDS-PAGE loading buffer is a harsh treatment that will denature all protein in the sample and restricts the sample to SDS-PAGE analysis only. This method may also strip excess protein off the affinity support that is nonspecifically bound to the matrix, and this material can interfere with analysis. Competitive analyte elution is much more specific for the bait–prey interaction, because it does not strip proteins that are nonspecifically bound to the affinity support. This method is non-denaturing; thus, it can elute a biologically functional protein complex, which could be useful for subsequent studies.

An alternative elution protocol is to use a step-wise gradient of increasing salt concentration or decreasing pH, which allows the selective elution of prey proteins while the bait remains immobilized. A gradient elution is not necessary once the critical salt concentration or pH has been optimized for efficient elution. These elution methods are also non-denaturing and can be informative in determining the relative interaction strength.

Gel Detection of Bait–prey Complex

Protein complexes contained in eluted samples can be visualized by SDS-PAGE and associated detection methods, including gel staining, Western blotting detection and S-35 radioisotopic detection. The final determination of interacting proteins often entails protein band isolation from a polyacrylamide gel, tryptic digestion of the isolated protein and mass spectrometric identification of digested peptides.

Pull-down Assay Controls

Pull-down assays entail multiple steps often using more than one cell lysate, and therefore each experiment must be properly controlled to demonstrate that the final results are not artifactual. Each experiment should analyze the lysate(s) both before and after being passed through the support to identify any nonspecific binding to the support. Each wash should also be analyzed to observe any eluted protein, and bait- and prey-free controls should also be used to confirm that there are no bait–prey interactions in the bait lysate and that the prey protein does not bind to the immobilized support.

Pull-down Methodologies

Homemade pull-down approaches for confirming or identifying protein–protein interactions are ubiquitous in contemporary scientific literature. The homemade pull-down assay represents a collection of reagents from multiple commercial vendors that cannot be validated together as a functional assembly except by extensive assay development by the researcher, and troubleshooting this combination of reagents can be tedious and time consuming. Commercial pull-down kits contain complete, validated sets of reagents specifically developed for performing pull-down assays. The buffers provided in each kit allow complete flexibility to determine the optimal conditions for isolating interacting proteins. The working solutions for washing and binding are physiologic in pH and ionic strength, providing a starting point from which specific buffer conditions for each unique interacting pair can be optimized. Many commercial kits also incorporate spin columns for efficient handling of small volumes of affinity support, complete retention of the affinity support during the pull-down assay and thorough washing of the protein complexes for minimal nonspecific protein pull-down, all of which are common sources of variability and high background using traditional pull-down assay formats.

In Vivo Techniques

Bimolecular Fluorescence Complementation

Bimolecular fluorescence complementation (BiFC) is a recent technique used in the investigation and direct visualization of protein–protein interactions (PPIs) and interaction between proteins and other macro-molecules that are essential for survival of cells. Identification of these relations specifies their working inside the cells in brief. Organisms with PPI networks include humans, worms, yeast, and plants.

BiFC is mainly used in genome-wide studies of PPIs; it is identified as an effective technique in uncovering new protein interactions, and providing novel information on functions of protein.

Principle of BiFC

The working principle of BiFC is based on the development of a fluorescent complex, as a result of the association of two segments of a fluorescent protein when they are in close proximity due to protein–protein interaction in the fragments, i.e., in BiFC, a fluorophore is divided into amino and carboxyl terminal ends. These ends are combined into two proteins. When these two proteins interact, both the segments re-associate resulting in re-formation of the fluorophore and fluorescence at the sites of interface.

Set of Rules to Design BiFC

The BiFC experiments – as it is based on the interaction between the fluorescent protein segments - takes place only under certain circumstances. BiFC analysis involves an assay that is fabricated in such a way that it satisfies all the parameters affecting the association of the fluorescent protein segments.

BiFC Calibration

BiFC calibration is carried out in four stages as follows:

1. Selection of appropriate fluorophore that works finely as BiFC synthesis partners. Examples of fluorophores are Venus and yellow fluorescent protein (YFP).

2. Labeling of proteins in which the BiFC segments are bonded to the amino- or carboxyl-terminals of the selected protein.

3. Determination of transfection circumstances before the testing of multiple mutants.

4. Determination of changes that occur in the localization of proteins due to addition of BiFC segments.

Plating and Cell Transfection

In this process, the cells are initially plated in a glass bottom plate and in two wells of a 6-well plate and then they are allowed to settle overnight at room temperature. Then the cell DNAs are prepared for transfection process. After completion of the transfection process, the combination is divided equally and then incubated at room temperature for a few hours. Then the transfection media is removed and incubated once again at room condition.

Preparation of Cells for Imaging

In this process, the cells are initially analyzed using an epifluorescent microscope in order to check their fluorescence. Then, the preparation of cells is done by addition of paraformaldehyde after which the cells are lysed to obtain the image.

Imaging of Cells

The intensity of fluorescence is determined for every single cell. Selective analysis of BiFC signals is achieved by performing transfection along with CFP. Imaging is done using a confocal microscope.

Fluorescence Quantitation

Measurement of the intensity of fluorescence is accomplished using several imaging software.

Analysis of Data

The efficiencies of fluorescence complementation are determined by the value obtained by division of the intensities of fluorescence complementation and the intensities of whole fluorescent protein in each cell.

Design of Multicolor BiFC Experiment

Multicolor BiFC experiment involves the combination of substitutional partners and the segments of fluorescent proteins that yield compounds with different color bands. Sequential imaging is achieved by visualization of the compounds involving distinct emission and excitation wavelengths. Imaging of the compounds is done alternatively to prevent re-localization of either compound based on the lag between the imaging of the compounds.

The fluorescence intensities should be maintained at constant magnitude in order to prevent development of variation in the distributions caused by variation in the signal-to-background ratio. The disturbances between two fluorophores can also be rectified by imaging the cells that represent only one group of proteins.

Multicolor BiFC was first established in mammalian cells in studies that used various combinations of YFP and CFP. BiFC contrives to examine the relative efficiencies of bFos interaction with bJun and bATF2.

The improvement of a multicolor BiFC method for plant cells helps in the investigation of simultaneous development of alternative calcium sensor/protein kinase complexes in planta.

Applications of the BiFC Assay

Applications of BiFC include the imaging of steady-state distribution of compounds developed by grouping of proteins in various types of cells and organisms. It also enables the imaging of interactions between multiple proteins in a single cell in addition to alterations in the covalent proteins that provide additional information about the biological processes of protein compounds that are recently encountered.

It is also helpful in the identification of various protein segments that can be combined by the interaction between the proteins that are bonded to the segments that include β-galactosidase, ubiquitin, and dihydrofolate reductase. BiFC experiments are also helpful in the determination of topology of membrane proteins and high input and output inspection of small molecular effects on the protein compounds.

Yeast Two-Hybrid

The yeast 2-hybrid (Y2H) assay is a well-established technique to detect protein-protein interactions. This is an extremely powerful tool for researchers and is often used alongside one or two other methods to examine the multitude of interactions that take place in cells. The assay is straightforward to perform and generates high quality results in a short amount of time.

Principles

1. Y2H assay relies on the expression of a reporter gene (such as lacZ or GFP), which is activated by the binding of a particular transcription factor.

2. The transcription factor is comprised of a DNA-binding domain (BD) and an activation domain (AD).

3. The query protein of interest fused with the BD is known as the Bait, and the protein library fused with the AD is referred to as the Prey.

4. In order to activate the reporter gene expression, a transcriptional unit must be present at the gene locus, which is only possible if Bait and Prey interact.

Theory

Expression of a reporter gene requires the binding of a transcription factor, which normally consists of two functionally and structurally independent domains: DNA-binding (DB) and activation (AD) domains. The DB domain binds to the particular DNA sequence upstream of the reporter gene, while the AD domain activates reporter gene expression.

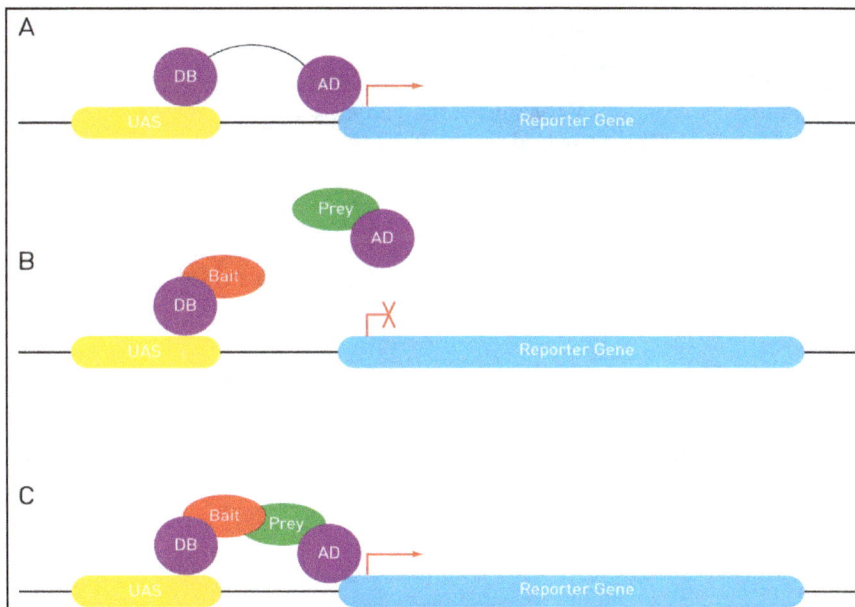

Y2H Principles.

Since AD and DB domains are functionally and structurally independent, they can be fused to two separate proteins. The protein that is fused to the DB domain is called the Bait, while the one fused to the AD domain is referred to as the Prey.

In the absence of Bait-Prey interaction, the AD domain is unable to localize to the reporter gene to drive gene expression. However, when Bait and Prey interact, the DB domain binds to the DNA localizing the AD domain upstream of the reporter gene, leading to the expression of reporter gene.

Yeast 2-Hybrid Variations

Several variations of the yeast 2-hybrid assay have been developed in order to study different protein interactions.

Yeast 1-Hybrid

The Yeast 1-hybrid assay examines protein-DNA interactions.

A query protein is directly fused with the AD domain and expressed in yeast strains harboring various target DNA sequences upstream of the reporter gene. Thus, if the query binds to a particular target sequence, the associated AD domain will activate reporter gene expression.

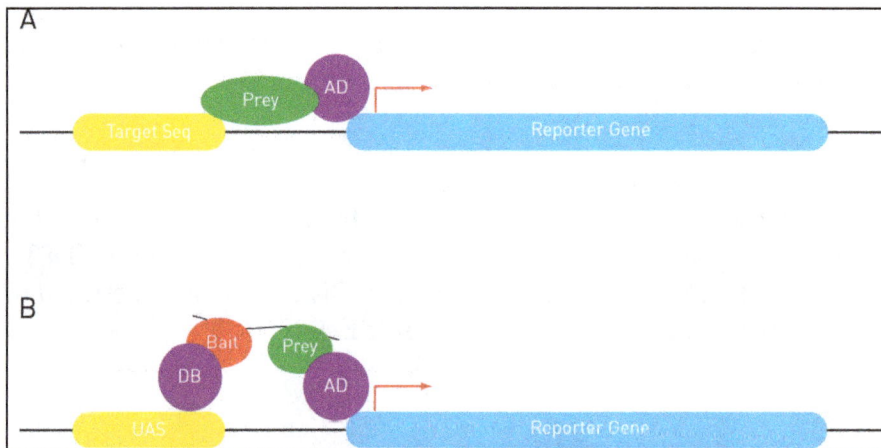

Variations of the Y2h Assay.
(A) Yeast 1-hybrid assay. (B) Yeast 3-hybrid assay.

Yeast 3-Hybrid

Yeast 3-hybrid assays study protein interactions that are mediated by a third component, such as an RNA molecule or another protein.

In this instance, Bait and Prey do not directly interact with each other. Instead, they bind to, for example, an RNA molecule, albeit with different sequence specificity. Therefore, only in the presence of the particular RNA molecule would Bait and Prey be able to interact and drive reporter gene expression.

Split Ubiquitin Yeast 2-Hybrid

The split ubiquitin assay identifies non-soluble membrane protein interactions.

Instead of fusing to AD and DB domains, query proteins are fused with one of the following two ubiquitin moieties: C-terminal ubiquitin moiety (Cub) and N-terminal ubiquitin moiety (Nub). The Cub is also fused with a transcription factor that can be cleaved by the specific protease.

The interaction of the query proteins allows the reconstitution of the ubiquitin, which in turn, is recognized by the protease. Cleavage of the protease will release the transcription factors to the nuclear, thereby activating the reporter gene expression.

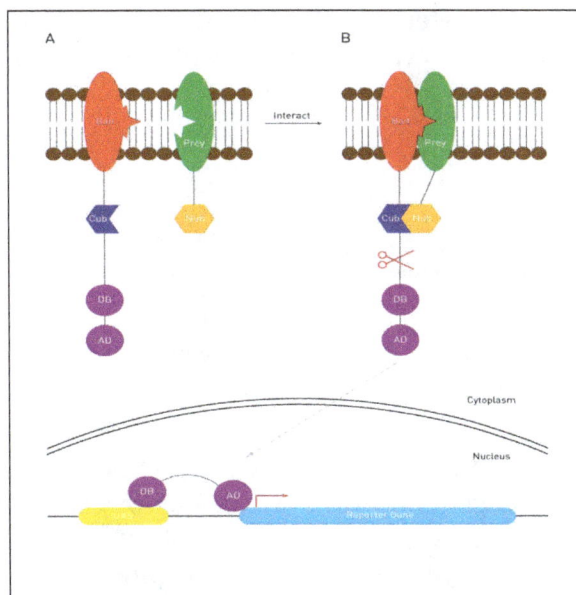

Split Ubiquitin Y2h.

Advantages and Limitations

Yeast 2-hybrid is a powerful technique to identify protein interactions because of its straightforward methodology and fast turnaround time. Therefore, the throughput of the technique can be scaled up significantly to screen the entire proteome. In addition, the technique has been adapted in other model organisms to study organism specific interactions.

It is important to recognize the limitations of yeast 2-hybrid assays:

1. The assay can produce a high level of false positive and negative interactions. This is an important reason to validate any interactions using other techniques such as co-immunoprecipitation.

2. Interaction must occur in the nucleus of the cell in order for the reporter gene to be activated. Proteins that are localized to other cellular compartments may not produce a positive interaction, even if they interact directly. This can be overcome by using a split ubiquitin 2-hybrid system or by performing the assay in bacteria, which do not possess a nucleus.

3. Overexpression of recombinant fusion protein, which happens in most yeast 2-hybrid experiments, could produce spurious interaction data. In addition, the fusion of AD/DB domains to query proteins may affect query protein function in vivo.

4. Query proteins may not be correctly expressed, folded, or modified when expressed in yeast. Therefore, it is important to confirm that query proteins are functional before deriving interaction data from the assay.

Working with Libraries

The availability of the Y2H Bait library collection has drastically elevated the throughput of the assay. A single Bait strain can be systematically crossed to an ordered array of yeast strains expressing individual Prey proteins, allowing proteome-wide screening and identification of novel protein interactions.

Crosslinking Protein Interaction Analysis

When two or more proteins have specific affinity for one another that causes them to come together in biological systems, bioconjugation technology can provide the means for investigating those interactions. Most in vivo protein–protein binding is transient and occurs only briefly to facilitate signaling or metabolic function. Capturing or freezing these momentary contacts to study which proteins are involved and how they interact is a significant goal of proteomics research today. Crosslinking reagents provide the means for capturing protein–protein complexes by covalently binding them together as they interact. The rapid reactivity of the common functional groups on crosslinkers allows even transient interactions to be frozen in place or weakly interacting molecules to be seized in a complex stable enough for isolation and characterization.

Chemical Crosslinkers

Crosslinking reagents covalently link together interacting proteins, domains or peptides by forming chemical bonds between specific amino acid functional groups on two or more biomolecules that occur in close proximity because of their interaction. Commercially available crosslinking reagents have a wide range of characteristics, including:

- Functional group specificity—the crosslinker molecule carries reactive moieties that target amines, sulfhydryls, carboxyls, carbonyls or hydroxyls.

- Homobifunctional or heterobifunctional—molecules are available with identical reactive moieties on both ends (termed homobifunctional molecules; the upper molecule (A/A) shown below, e.g., DSS) that crosslink identical residues, while each reactive group on heterobifunctional crosslinkers targets different functional groups on separate proteins for greater variability or specificity, as shown in the lower molecule below (A/B, e.g., SMCC).

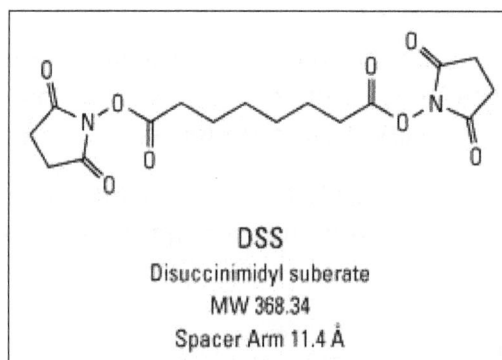

DSS
Disuccinimidyl suberate
MW 368.34
Spacer Arm 11.4 Å

Chemical structure of DSS.

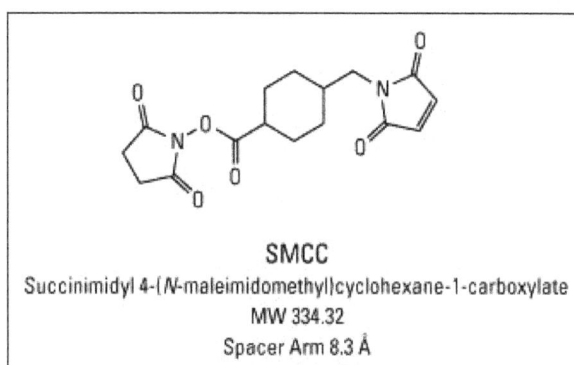

SMCC

Succinimidyl 4-(*N*-maleimidomethyl)cyclohexane-1-carboxylate

MW 334.32

Spacer Arm 8.3 Å

Chemical structure of SMCC.

- Variable spacer arm length—the reactive groups are spatially separated by the crosslinker molecule structure, which allows the crosslinking of amino acids that are varying distances apart, as shown below with Sulfo-EGS versus BS3. Zero-length crosslinkers are also available, which crosslink two amino acid residues without leaving any part of the crosslinker molecule remaining in the interaction after the reaction is completed.

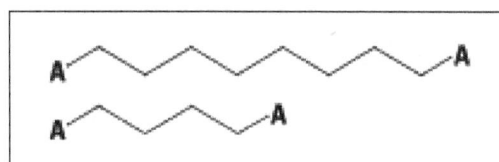

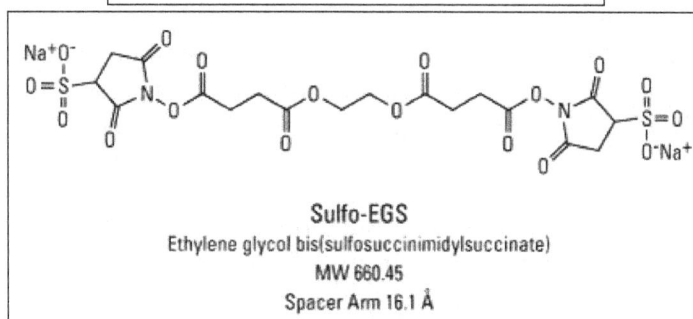

Sulfo-EGS

Ethylene glycol bis(sulfosuccinimidylsuccinate)

MW 660.45

Spacer Arm 16.1 Å

Chemical structure of Sulfo-EGS.

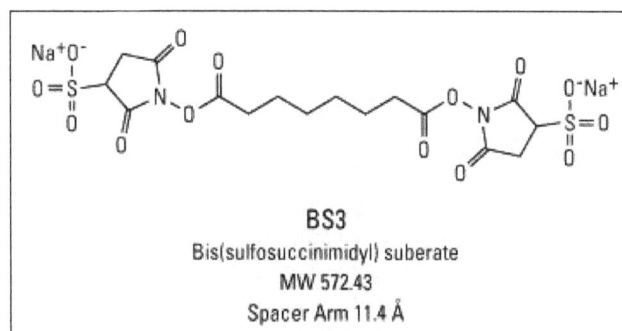

BS3

Bis(sulfosuccinimidyl) suberate

MW 572.43

Spacer Arm 11.4 Å

Chemical structure of BS3.

- Cleavable or non-cleavable—the crosslinker molecule can also be designed to include cleavable elements, such as esters or disulfide bonds (diagrammed below, e.g., DSP), to reverse or break the linkage by the addition of hydroxylamine or reducing agents, respectively.

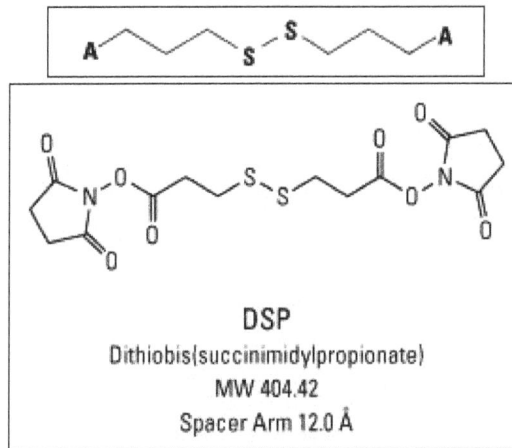

Chemical structure of DSP.

- Water-soluble or -insoluble—crosslinkers can be hydrophobic to allow passage into hydrophobic protein domains (DSS) or through the cell membrane or hydrophilic to limit crosslinking to aqueous compartments (BS3).

In Vivo and in Vitro Crosslinking

In Vivo Crosslinking

Besides the transient and sometimes tentative nature of some protein–protein interactions, the formation of these complexes can change in response to any number of stimuli, including changes in pH, temperature and osmolarity, and either the lack of a specific protein or co-factor or the introduction of a protein with which the protein(s) do not normally interact.

The benefit of in vivo crosslinking is that the protein–protein interaction can be captured in its native environment, which limits the risk of false positive interactions or the loss of complex stability during cell lysis. For in vivo crosslinking, hydrophobic, lipid-soluble crosslinkers are expected to be used if the target protein is within or across cell membranes, while hydrophilic, water-soluble crosslinkers can be used to crosslink cell surface proteins, such as receptor–ligand complexes. This representative data provides an example of various reagents used for in vivo crosslinking.

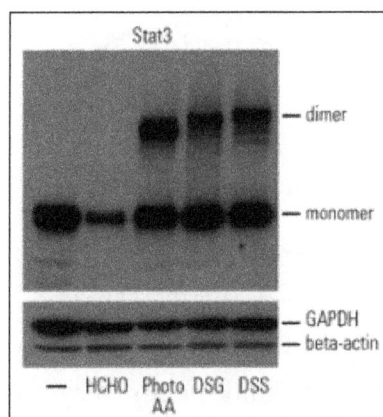

Comparison of several in vivo crosslinking methods.

Due to the high concentration of proteins in cells, crosslinkers with shorter spacer arms are usually recommended for in vivo crosslinking approaches to increase the specificity of conjugating actual interacting proteins as opposed to proteins that just happen to be in close proximity to each other during incubation with the crosslinker.

Although in vivo crosslinking can yield physiologically relevant, stably-crosslinked complexes for analysis, optimizing this approach can be difficult, as the reaction conditions cannot be tightly controlled and crosslinkers react with a wide array of proteins that all present functional groups against which crosslinkers specifically react.

In Vitro Crosslinking

In vitro crosslinking can better target specific crosslinking events, because more reaction conditions can be tightly controlled, including the pH, temperature, concentration of reactants and purity of the target protein(s). The ability to control all aspects of a conjugation experiment results in better analysis due to greater resolution of protein–protein interactions. Additionally, in vitro methods of conjugation allow researchers to modify interacting proteins, such as adding polyethylene glycol groups (PEGylation), blocking sulfhydryls or converting amines to sulfhydryls. Also, a greater variety of crosslinking reagents, both hydrophobic and hydrophilic, are available for in vitro applications. This representative data was produced using the amine-reactive crosslinking reagents, DSS, BS3, and DSSO.

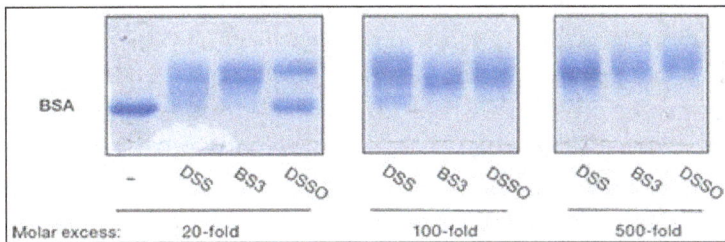

Comparison of BSA crosslinking efficiency by SDS-PAGE. Different crosslinkers were incubated with BSA at molar excess of crosslinker to protein (e.g., 20-, 100- or 500-fold). Crosslinking efficiency is shown by decreased mobility by SDS-PAGE and varied by crosslinker type, solubility and concentration.

Obviously, the disadvantage of using in vitro methods to conjugate proteins is the lack of physiological conditions. Additionally, rupturing and solubilizing membranes can disrupt protein–protein and protein–membrane interactions.

Because a myriad of crosslinking reagents are commercially available for many different applications, the key determinant in deciding to use in vivo or in vitro crosslinking is the target protein, specifically in term of its:

- Cellular location—in vivo crosslinking would benefit protein targets embedded in the cell membrane, while cytoplasmic proteins could be crosslinked by either method, depending on the next determinant.

- Interaction stability—weak protein–protein interactions may be lost during in vitro crosslinking due to cell lysis and potential competition with other proteins, while stable interactions may be strong enough to withstand these forces.

General Reaction Conditions

Choosing the Appropriate Crosslinker

Correct identification of protein-protein interactions first requires the selection of the best crosslinker to use. Because there are multiple amino acid functional groups that may react with different crosslinkers, an empirical strategy of screening multiple types of crosslinkers should first be performed to identify the target protein conjugate. The crosslinkers tested may vary in:

- Hydrophobicity.

- Reactive groups.

- Homo- vs. heterobifunctionality.

- Spacer arm length.

Once the target interaction is detected by any of the methods listed below, then the protocol can be fine-tuned to optimize detection by adjusting crosslinker concentration, pH and other reaction conditions.

Sample Preparation

The starting protein concentration or number of cells should be empirically determined for in vitro and in vivo crosslinking protocols, respectively. For in vitro crosslinking, the protein solution should be prepared in a nonreactive buffer, such as phosphate-buffered saline (PBS), which has the proper pH for the specific crosslinker. For in vivo crosslinking applications, cells should be in the exponential phase of growth and at a subconfluent density during the crosslinking procedure. To avoid the possibility of culture media reacting with the crosslinker, the media can be replaced with PBS through a series of cell washes.

Reaction Conditions

Crosslinkers should be prepared as per the manufacturer's instructions; hydrophobic crosslinkers are first dissolved in the appropriate solvent, such as methanol or acetone. The optimum amount of reagent to add also depends on the crosslinker, but usually a 20- to 500-fold molar excess (relative to the lysate protein concentration) is appropriate. Ensure that pH of the reaction buffer is favorable for the crosslinker. Most amine-reactive crosslinkers require alkaline pH for activity.

The crosslinking reaction time may also be important, depending upon the experiment and crosslinker being used. While 30 minutes is a good incubation time to start with, multiple experiments can be performed concurrently to test other lengths of time to determine the optimal time of incubation with the specific crosslinker. Long incubation periods should generally be avoided, not only because it may cause formation of large, crosslinked protein aggregates, but also because the crosslinker may lose stability. In cases where extended incubation periods are required, though, fresh crosslinker can be added at specific time points throughout the procedure to maintain the proper molar ratio of reagent and maximize the formation of the target product. The formation of aggregates due to extensive crosslinking, though, should also be considered in determining the optimal reaction time.

Quenching the Reaction

With most amine-reactive crosslinkers used for protein–protein interaction analysis, the reaction can be halted at the desired time by adding excess nucleophile, such as Tris or glycine, which out-competes the lysate proteins for reaction with the crosslinker. The crosslinked product can then be purified through multiple approaches, including precipitation, chromatography, dialysis or ultrafiltration.

A rapid method that combines quenching the reaction and denaturing the proteins in preparation for gel electrophoresis is to add sodium dodecyl sulfate–polyacrylamide gel electrophoresis (SDS-PAGE) buffer, which contains both Tris and 2-Mercatpoethanol, and then boil the solution for 5 minutes. The sample can then be directly analyzed by gel electrophoresis.

Protein–protein Interaction Analysis

Crosslinking is typically used to capture and stabilize transient or labile interactions so that they can be further isolated and analyzed by downstream methods such as electrophoresis, staining, western blot, immunoprecipitation or co-immunoprecipitation and mass spectrometry.

Western Blot

When two proteins are covalently crosslinked, the gel migration patterns of both proteins shift in relation to the uncrosslinked proteins. Therefore, if antibodies that detect each target protein are available, the most straightforward method to detect the shift of the interacting proteins is by SDS-PAGE and western blot analysis.

Immunoprecipitation and Co-immunopreciptation (co-IP)

Both immunoprecipitation (IP) and co-immunoprecipitation (co-IP) are methods to detect protein expression and protein–protein interactions, respectively, via affinity purification. Crosslinking is commonly performed in both applications, either alone or in combination with affinity binding, to immobilize antibody to the beaded support and or freeze weak antibody–antigen interactions to prevent sample loss during immune complex extraction. Crosslinking is also used to stabilize transient or weak protein–protein interactions prior to co-IP protocols. Following both approaches, samples are commonly analyzed by SDS-PAGE.

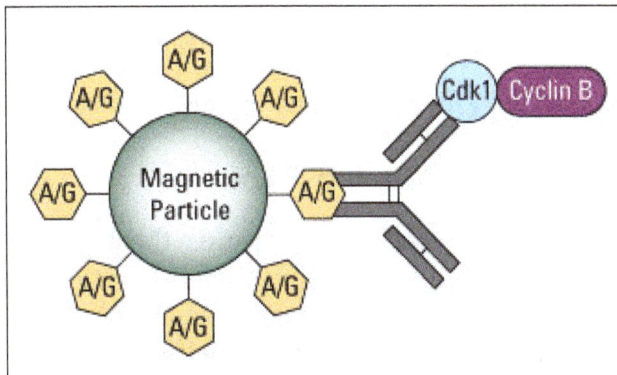

Co-immunoprecipitation of cyclin B and Cdk1.

Mass Spectrometry

When analysis by mass spectrometry (MS) is available, the peptide fragments that are crosslinked between interacting proteins can be identified by the change in mass resulting from the attached crosslinker molecule. In this approach, identical samples are crosslinked with either deuterated (heavy) or nondeuterated (light) crosslinkers. The crosslinked proteins are then pooled together and analyzed by MS to identify and quantify the heavy product based on its shift from the light product. This method also commonly employs SDS-PAGE as a first-stage purification step prior to digestion in preparation for MS analysis.

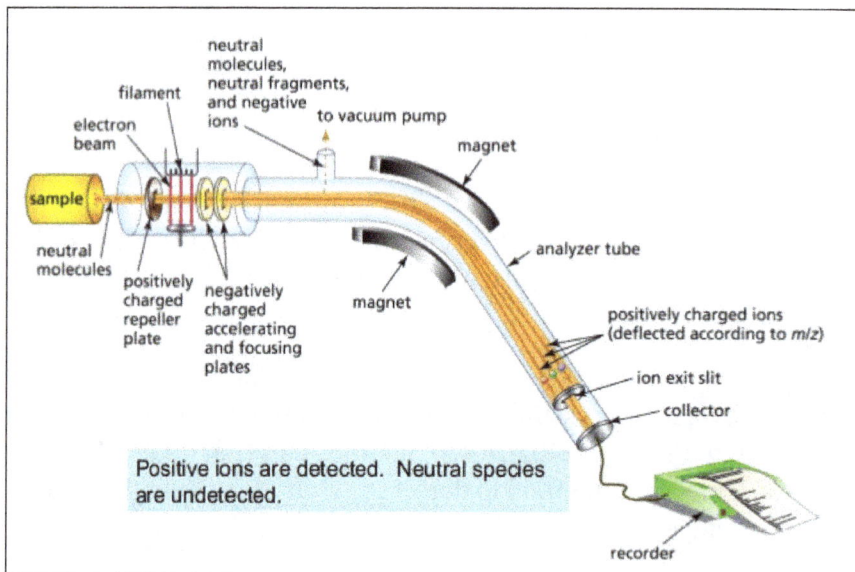

Interactome

An interactome network refers to all the protein-protein interactions which take place inside a cell. Such interactome maps have been created for yeast, worms, flies, and are beginning to be created for humans.

In yeast, the size of the interactome is estimated to be approximately 28,000 protein-protein interactions, while the number of interactions in humans is estimated to be around 650,000.

Different methods used to map and analyse the interactome are discussed below.

Fluorescence Cross-Correlation Spectroscopy

Fluorescence correlation spectroscopy or FCS measures the thermodynamic fluctuations in the fluorescence intensity of tagged proteins. Using confocal microscope, the dynamics of how fluorescently-tagged protein molecules move though a small focal volume are quantified. The diffusion properties of the tagged proteins depend on their concentration and interactions with other molecules. In fluorescence cross correlation microscopy, two fluorescently labeled proteins are visualised. If these proteins are associated with each other, they move in a synchronized manner. This leads to similar fluctuations in their fluorescent signals over time. Thus, quantitative FCCS can measure protein-protein interactions to generate quantitative interactome maps.

Bimolecular Complementation Methods

Protein-protein Complementation Assay (PCA)

This method is also used to study protein-protein interactions. In this method, a fluorescent protein is split in to two, and then fused to the N-or C-terminals of two potential interacting proteins. If the proteins interact, the two units combine, leading to fluorescence which can then be visualised.

BiFC

This method is a variant of PCA where GFP is split into two parts and fused to potentially interacting peptides. The interaction of peptides leads to assembly of functional GFP molecules which then fluoresce.

BiLC

This is another variant of PCA where, instead of fluorescent proteins, luciferase is used. Again the luciferase is fragmented into two to visualise interactions. An important property of this method is that the association of the luciferase fragments is irreversible. Thus, this method can be used to study both interaction and dissociation of proteins inside a cell.

FRET-Based Methods

FRET or Forster Energy Resonance Transfer is a method based upon the distance between two fluorophores. In this method, if located close enough, an excited fluorophore molecule (donor) transfers its energy to another fluorophore molecule (acceptor). This leads to a fluorescence emission in the acceptor molecule. The measurements of the excitation and emission wavelengths can be used to measure interactions between two molecules as FRET can occur only when the donor or acceptor molecules are in very close proximity or in direct interaction with each other. The limitation of this method is that studying protein-protein interaction to form a interactome map requires tagging the proteins of interest with the appropriate acceptor and donor fluorophores. This requires creating genetic fusion proteins.

BRET-Based Methods

BRET or bioluminescence-based resonance energy transfer involves oxidation of a substrate by luciferase enzyme. The oxidized substrate then emits light at 395 nm which is captured by an acceptor GFP molecule leading to GFP fluorescence. As there is no need for external light source in BRET, there is less background noise, autoflourescence, and light scattering.

Luciferase-Based Co-Immunoprecipitation Methods

This method is also used to detect protein-protein interactions to generate the interactome map. This method is less laborious and time-consuming than the conventional co-immunoprecipitation methods. In this method, the acceptor and donor proteins are fused with FLAG and Renilla constructs. Any interactions or associations can be detected using the luciferase enzymatic assay.

Using this method, a large number of donor/acceptor or interacting partners can be detected. This is in contrast to the conventional co-immunoprecipitation methods where only a single interaction can be determined. This advantage is critical while constructing interactome maps which involve thousands of interactions. However, one disadvantage of this method is that the FLAG-tagged protein cannot be quantified which may generate false negative results.

Proximity Ligation Assays

In this method, antibodies are attached to short single-stranded DNA oligonucleotides or PLA probes. These antibodies then bind to protein inside cells. The proximity or interaction of two proteins can lead to ligation of DNA molecules which can then be amplified using a polymerase chain reaction. The fluorescent probes then bind to these amplified DNA regions which then acts as a marker to show interacting proteins.

Thus, different methods can be used to map interactions in a cell which are then used to form the interactome.

References

- Protein-protein-interaction-analysis, pierce-protein-methods, protein-biology, life-science: thermofisher.com, Retrieved 8 April, 2019

- Immunoprecipitation: profacgen.com, Retrieved 14 January, 2019

- https://www.mybiosource.com/learn/assay-learning-center/coimmunoprecipitation/, Retrieved 9 June, 2019

- Coimmunoprecipitation: mybiosource.com, Retrieved 6 May, 2019

- Interaction-of-proteins, science: whatisbiotechnology.org, Retrieved 16 March, 2019

- Pull-down, pierce-protein-methods, protein-biology, life-science: thermofisher.com, Retrieved 11 February, 2019

- Lowman HB, Clackson T (2004). "1.3". Phage display: a practical approach. Oxford [Oxfordshire]: Oxford University Press. pp. 10–11. ISBN 978-0-19-963873-4

- Bimolecular-Fluorescence-Complementation, life-sciences: news-medical.net, Retrieved 20 August, 2019

- Yeast-2-hybrid: singerinstruments.com, Retrieved 2 July, 2019

- Crosslinking-protein-interaction-analysis, pierce-protein-methods, protein-biology, life-science: thermofisher.com, Retrieved 12 May, 2019

- What-is-the-Interactome, life-sciences: news-medical.net, Retrieved 20 March, 2019

Chapter 4

Protein Post-Translational Modifications

Post- translational modification is the enzymatic and covalent modification of proteins that follows protein biosynthesis. Proteins are synthesized by ribosomes and then undergo post-translational modification in order to form mature protein product. This chapter has been carefully written to provide an easy understanding of the varied kinds of post-translational modifications such as acetylation, methylation and ubiquitination.

Protein post-translational modifications (PTMs) increase the functional diversity of the proteome by the covalent addition of functional groups or proteins, proteolytic cleavage of regulatory subunits, or degradation of entire proteins. These modifications include phosphorylation, glycosylation, ubiquitination, nitrosylation, methylation, acetylation, lipidation and proteolysis and influence almost all aspects of normal cell biology and pathogenesis. Therefore, identifying and understanding PTMs is critical in the study of cell biology and disease treatment and prevention.

Within the last few decades, scientists have discovered that the human proteome is vastly more complex than the human genome. While it is estimated that the human genome comprises between 20,000 and 25,000 genes, the total number of proteins in the human proteome is estimated at over 1 million. These estimations demonstrate that single genes encode multiple proteins. Genomic recombination, transcription initiation at alternative promoters, differential transcription termination, and alternative splicing of the transcript are mechanisms that generate different mRNA transcripts from a single gene.

The increase in complexity from the level of the genome to the proteome is further facilitated by protein post-translational modifications (PTMs). PTMs are chemical modifications that play a key role in functional proteomic because they regulate activity, localization, and interaction with other cellular molecules such as proteins, nucleic acids, lipids and cofactors.

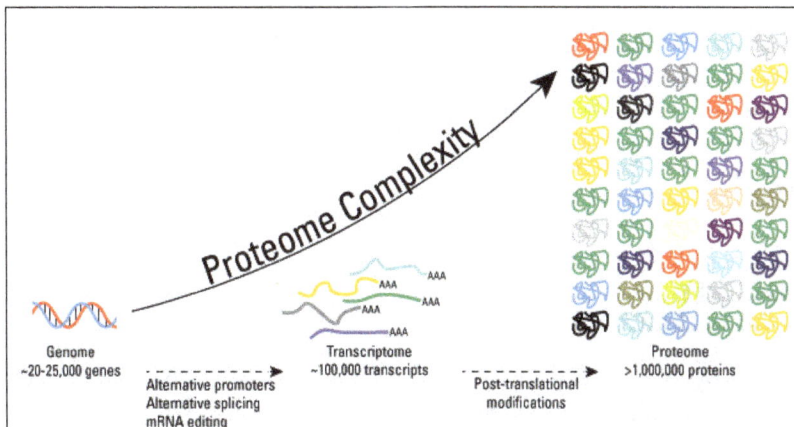

Post-translational modifications are key mechanisms to increase proteomic diversity. While the genome comprises 20,000 to 25,000 genes, the proteome is estimated to encompass over 1

million proteins. Changes at the transcriptional and mRNA levels increase the size of the transcriptome relative to the genome, and the myriad of different post-translational modifications exponentially increases the complexity of the proteome relative to both the transcriptome and genome.

Additionally, the human proteome is dynamic and changes in response to a legion of stimuli, and post-translational modifications are commonly employed to regulate cellular activity. PTMs occur at distinct amino acid side chains or peptide linkages, and they are most often mediated by enzymatic activity. Indeed, it is estimated that 5% of the proteome comprises enzymes that perform more than 200 types of post-translational modifications. These enzymes include kinases, phosphatases, transferases and ligases, which add or remove functional groups, proteins, lipids or sugars to or from amino acid side chains; and proteases, which cleave peptide bonds to remove specific sequences or regulatory subunits. Many proteins can also modify themselves using autocatalytic domains, such as autokinase and autoprotolytic domains.

Post-translational modification can occur at any step in the "life cycle" of a protein. For example, many proteins are modified shortly after translation is completed to mediate proper protein folding or stability or to direct the nascent protein to distinct cellular compartments (e.g., nucleus, membrane). Other modifications occur after folding and localization are completed to activate or inactivate catalytic activity or to otherwise influence the biological activity of the protein. Proteins are also covalently linked to tags that target a protein for degradation. Besides single modifications, proteins are often modified through a combination of post-translational cleavage and the addition of functional groups through a step-wise mechanism of protein maturation or activation.

Protein PTMs can also be reversible depending on the nature of the modification. For example, kinases phosphorylate proteins at specific amino acid side chains, which is a common method of catalytic activation or inactivation. Conversely, phosphatases hydrolyze the phosphate group to remove it from the protein and reverse the biological activity. Proteolytic cleavage of peptide bonds is a thermodynamically favorable reaction and therefore permanently removes peptide sequences or regulatory domains.

Consequently, the analysis of proteins and their post-translational modifications is particularly important for the study of heart disease, cancer, neurodegenerative diseases and diabetes. The characterization of PTMs, although challenging, provides invaluable insight into the cellular functions underlying etiological processes. Technically, the main challenges to studying post-translationally modified proteins are the development of specific detection and purification methods. Fortunately, these technical obstacles are being overcome with a variety of new and refined proteomics technologies.

Ubiquitination

Ubiquitination, also known as ubiquitylation, is an enzymatic process that involves the bonding of an ubiquitin protein to a substrate protein. This has sometimes been referred to as the molecular "kiss of death" for a protein, as the substrate usually becomes inactivated and is tagged for degradation by the proteasome through the attachment of the ubiquitin molecule.

Post-Translation Modification Enzymes

The process of ubiquitination in regulated by three main types of enzymes to take place in entirety. These include ubiquitin-activating enzymes (E1), ubiquitin conjugating enzymes (E2) and ubiquitin ligases (E3). Each of these enzyme types has an important role to play in ubiquitination and the labeling of proteins to be degraded by the proteasome, which are considered in more detail below.

The first step involves the activation of ubiquitin by the E1 enzyme, which occurs prior to its attachment to the amino acid cysteine, the active site. Energy in the form of ATP is required in order for the ubiquitin molecule to be transferred to the active site and produce an intermediate substance known as ubiquitin-adenylate.

Following this, the ubiquitin-conjugating enzyme (E2) plays its role to bring the two molecules together, by transferring the ubiquitin from E1 to the active cysteine site. The E2 enzyme has a particular structure that allows it to bond to both the ubiquitin and E1 molecules and allow this step to occur.

Finally, the ubiquitin protein ligase (E3) is required to recognise and bind the target substrate, subsequently labeling it with the small ubiquitin molecule. This usually occurs by way of an isopeptide bond connecting the last amino acid, glycine 76, of the ubiquitin molecule to a lysine on the substrate protein.

This enzymatic process is then repeated to form a small chain with several ubiquitin molecules, marking the protein for degradation in the proteasome.

Cases of Non-Degradation

Although the labeling of a protein via ubiquitination largely results in the degradation of the protein, there are some cases in which it may not prove fatal.

For example, when a single ubiquitin molecule is bound to a protein without forming a chain of molecules, which is known as mono-ubiquitination, the result can differ significantly. It is common for the protein to instead notice an alteration in function or it may be degraded via lysosomes rather that in the proteasome. This can also occur to some proteins that have undergone poly-ubiquitination, although it is less common.

Additionally, the process of ubiquitination can be reversed through the action of deubiquitinase enzymes, which break the bond between the ubiquitin molecule and the substrate protein.

Phosphorylation

Reversible protein phosphorylation, principally on serine, threonine or tyrosine residues, is one of the most important and well-studied post-translational modifications. Phosphorylation plays critical roles in the regulation of many cellular processes including cell cycle, growth, apoptosis and signal transduction pathways.

Phosphorylation is the most common mechanism of regulating protein function and transmitting signals throughout the cell. While phosphorylation has been observed in bacterial proteins, it is considerably more pervasive in eukaryotic cells. It is estimated that one-third of the proteins in the human proteome are substrates for phosphorylation at some point. Indeed, phosphoproteomics has been established as a branch of proteomics that focuses solely on the identification and characterization of phosphorylated proteins.

Mechanism of Phosphorylation

While phosphorylation is a prevalent post-translational modification (PTM) for regulating protein function, it only occurs at the side chains of three amino acids, serine, threonine and tyrosine, in eukaryotic cells. These amino acids have a nucleophilic (−OH) group that attacks the terminal phosphate group (γ-PO_3^{2-}) on the universal phosphoryl donor adenosine triphosphate (ATP), resulting in the transfer of the phosphate group to the amino acid side chain. This transfer is facilitated by magnesium (Mg^{2+}), which chelates the γ- and β-phosphate groups to lower the threshold for phosphoryl transfer to the nucleophilic (−OH) group. This reaction is unidirectional because of the large amount of free energy that is released when the phosphate–phosphate bond in ATP is broken to form adenosine diphosphate (ADP).

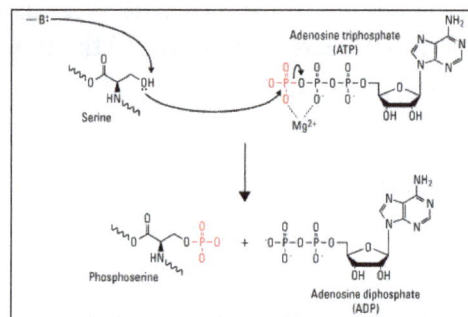

Diagram of serine phosphorylation. Enzyme-catalyzed proton transfer from the (−OH) group on serine stimulates the nucleophilic attack of the γ-phosphate group on ATP, resulting in transfer of the phosphate group to serine to form phosphoserine and ADP. (−B:) indicates the enzyme base that initiates proton transfer.

For a large subset of proteins, phosphorylation is tightly associated with protein activity and is a key point of protein function regulation. Phosphorylation regulates protein function and cell signaling by causing conformational changes in the phosphorylated protein. These changes can affect the protein in two ways. First, conformational changes regulate the catalytic activity of the protein. Thus, a protein can be either activated or inactivated by phosphorylation. Second, phosphorylated proteins recruit neighboring proteins that have structurally conserved domains that recognize and bind to phosphomotifs. These domains show specificity for distinct amino acids. For example, Src homology 2 (SH2) and phosphotyrosine binding (PTB) domains show specificity for phosphotyrosine (pY), although distinctions in these two structures give each domain specificity for distinct phosphotyrosine motifs. Phosphoserine (pS) recognition domains include MH2 and the WW domain, while phosphothreonine (pT) is recognized by forkhead-associated (FHA) domains. The ability of phosphoproteins to recruit other proteins is critical for signal transduction, in which downstream effector proteins are recruited to phosphorylated signaling proteins.

Protein phosphorylation is a reversible PTM that is mediated by kinases and phosphatases, which phosphorylate and dephosphorylate substrates, respectively. These two families of enzymes facilitate the dynamic nature of phosphorylated proteins in a cell. Indeed, the size of the phosphoproteome in a given cell is dependent upon the temporal and spatial balance of kinase and phosphatase concentrations in the cell and the catalytic efficiency of a particular phosphorylation site.

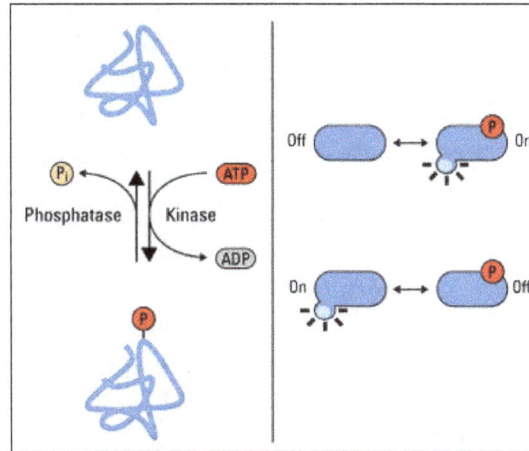

Phosphorylation is a reversible PTM that regulates protein function. Left panel: Protein kinases mediate phosphorylation at serine, threonine and tyrosine side chains, and phosphatases reverse protein phosphorylation by hydrolyzing the phosphate group. Right panel: Phosphorylation causes conformational changes in proteins that either activate (top) or inactivate (bottom) protein function.

Protein Kinases

Kinases are enzymes that facilitate phosphate group transfer to substrates. Greater than 500 kinases have been predicted in the human proteome; this subset of proteins comprises the human kinome. Substrates for kinase activity are diverse and include lipids, carbohydrates, nucleotides and proteins.

ATP is the cosubstrate for almost all protein kinases, although guanosine triphosphate is used by a small number of kinases. ATP is the ideal structure for the transfer of α-, β- or γ-phosphate groups for nucleotidyl-, pyrophosphoryl- or phosphoryltransfer, respectively. While the substrate specificity of kinases varies, the ATP-binding site is generally conserved.

Protein kinases are categorized into subfamilies that show specificity for distinct catalytic domains and include tyrosine kinases or serine/threonine kinases. Approximately 80% of the mammalian kinome comprises serine/threonine kinases, and >90% of the phosphoproteome consists of pS and pT. Indeed, studies have shown that the relative abundance ratio of pS:pT:pY in a cell is 1800:200:1. Although pY is not as prevalent as pS and pT, global tyrosine phosphorylation is at the forefront of biomedical research because of its relation to human disease via the dysregulation of receptor tyrosine kinases (RTKs).

Protein kinase substrate specificity is based not only on the target amino acid but also on consensus sequences that flank it. These consensus sequences allow some kinases to phosphorylate single proteins and others to phosphorylate multiple substrates (>300). Additionally, kinases can phosphorylate single or multiple amino acids on an individual protein if the kinase-specific consensus sequences are available.

Kinases have regulatory subunits that function as activating or autoinhibitory domains and have various regulatory substrates. Phosphorylation of these subunits is a common approach to regulating kinase activity. Most protein kinases are dephosphorylated and inactive in the basal state and are activated by phosphorylation. A small number of kinases are constitutively active and are made intrinsically inefficient, or inactive, when phosphorylated. Some kinases, such as Src, require a combination of phosphorylation and dephosphorylation to become active, indicating the high regulation of this proto-oncogene. Scaffolding and adaptor proteins can also influence kinase activity by regulating the spatial relationship between kinases and upstream regulators and downstream substrates.

The activity of specific kinases can be measured by incubating immunoprecipitates with substrates for specific kinases and ATP. Commercial kits are available for this type of assay and are designed to yield colorimetric, radiometric or fluorometric detection. While this type of assay shows the activity of specific kinases, as with kinase enrichment, it does not provide information on the proteins that the kinases modify or the role of endogenous phosphatase activity.

Signal Transduction Cascades

The reversibility of protein phosphorylation makes this type of PTM ideal for signal transduction, which allows cells to rapidly respond to intracellular or extracellular stimuli. Signal transduction cascades are characterized by one or more proteins physically sensing cues, either through ligand binding, cleavage or some other response, that then relay the signal to second messengers and signaling enzymes. In the case of phosphorylation, these receptors activate downstream kinases, which then phosphorylate and activate their cognate downstream substrates, including additional kinases, until the specific response is achieved. Signal transduction cascades can be linear, in which kinase A activates kinase B, which activates kinase C and so forth. Signaling pathways have also been discovered that amplify the initial signal; kinase A activates multiple kinases, which in turn activate additional kinases. With this type of signaling, a single molecule, such as a growth factor, can activate global cellular programs such as proliferation.

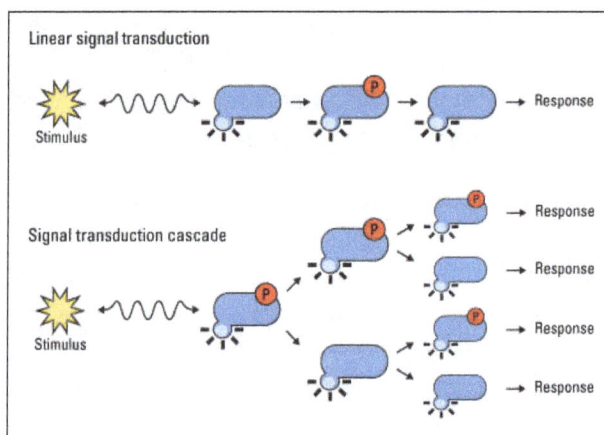

Signal transduction cascades amplify the signal output. External and internal stimuli induce a wide range of cellular responses through a series of second messengers and enzymes. Linear signal transduction pathways yield the sequential activation of a discrete number of downstream effectors, while other stimuli elicit signal cascades that amplify the initial stimulus for large-scale or global cellular responses.

Protein Phosphatases

The intensity and duration of phosphorylation-dependent signaling is regulated by three mechanisms:

- Removal of the activating ligand.

- Kinase or substrate proteolysis.

- Phosphatase-dependent dephosphorylation.

The human proteome is estimated to contain approximately 150 protein phosphatases, which show specificity for pS/pT and pY residues. While dephosphorylation is the end goal of these two groups of phosphatases, they do it through separate mechanisms. Serine/threonine phosphatases mediate the direct hydrolysis of the phosphorus atom of the phosphate group using a bimetallic (Fe/Zn) center, while tyrosine phosphatases form a covalent thiophosphoryl intermediate that facilitates removal of the tyrosine residue.

Methods of Detection

Because of the influence that phosphorylation has on biological processes in general, a huge emphasis has been placed on understanding the biological role of protein phosphorylation in the context of human disease. Small-scale protein phosphorylation is commonly performed to study the activity of a small number of proteins, while phosphoproteomic analyses are increasingly used to understand the global dynamics of phosphorylation of entire protein families. Current approaches to study protein phosphorylation include immunodetection, phosphoprotein or phosphopeptide enrichment, kinase activity assays and mass spectrometry. Because of the detrimental effect that phosphatases can have on the detection of phosphorylated proteins, broad-spectrum phosphatase inhibitors are commonly added to cell lysates in many phosphodetection strategies.

Immunodetection

Phospho-specific antibodies raised against specific phospho-epitopes on target proteins are a core tool for studying site-specific protein phosphorylation. On a more global scale, antibodies have been developed to detect the phosphorylation of specific amino acids (pS, pT, pY). Phospho-specific antibodies can be used for traditional western blotting, immunoprecipitation (IP), immunohistochemistry (IHC), ELISA, flow cytometry and, more recently, immobilization onto solid support arrays. Although phospho-specific antibodies are popular amongst the research community, only a small fraction of them are highly specific for their targets and many have issues with low sensitivity.

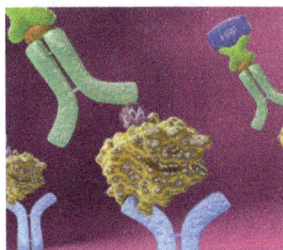

Phosphorylation-specific ELISA kits. Phosphorylation-specific Invitrogen ELISA kits are high-quality ELISA kits, designed for researchers studying intracellular proteins involved in signaling pathways.

These assay kits are designed to deliver accurate, sensitive, and fast protein quantitation of total and phosphorylated, modified, or cleavage site-specific proteins in a broad range of sample types.

In the following example, western blot analysis was used to detect the protein p38 MAPK, a serine/threonine kinase that plays an important role in signal transduction, contributing to the regulation of many cellular processes including cell differentiation and inflammation.

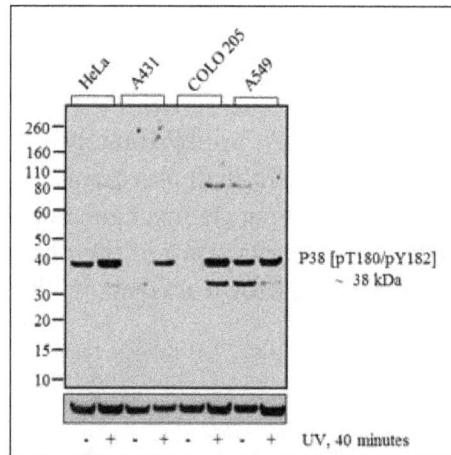

Detection of p38 MAP protein. Western blot analysis of p38 [pT180]/[pY182] was performed by loading 20 μg of HeLa (lane 1), HeLa exposed for 40 minutes with UV (lane 2), A431 (lane 3), A431 exposed for 40 minutes with UV (lane 4), COLO 205 (lane 5), COLO 205 exposed for 40 minutes with UV (lane 6), A549 (lane 7) and A549 exposed for 40 minutes with UV (lane 8) cell lysate using Invitrogen NuPAGE 4-12% Bis-Tris Gel (Cat. No. NP0322BOX), XCell SureLock Electrophoresis System (EI0002), Novex Sharp Pre-Stained Protein Standard (LC5800), and iBlot Dry Blotting System (IB21001). Proteins were transferred to a nitrocellulose membrane and blocked with 5% skim milk for 1 hour at room temperature. p38 [pT180]/[pY182] was detected at ~38 kDa using Invitrogen p38 [pT180]/[pY182] rabbit polyclonal antibody (44684G) at 1:1,000 in 5% skim milk at 4 °C overnight on a rocking platform. Invitrogen goat anti-rabbit IgG–HRP secondary antibody (G21234) at 1:5,000 dilution was used and chemiluminescent detection was performed using Invitrogen Novex ECL Chemiluminescent Substrate Reagent Kit.

Enrichment

Key dynamic changes in protein phosphorylation occur on low abundance proteins, which require enrichment prior to proteomic analysis. These strategies include metal oxide affinity chromatography (MOAC) and immobilized metal affinity chromatography (IMAC), which use metal-ligand complexes to capture phosphate groups on pS, pT and pY. MOAC is most commonly performed with TiO2-chelated resins to form bidentate complexes with phosphates, while IMAC employs Fe-chelated support to form tri- or tetradentate complexes with phosphates. Commercial kits are also available that have proprietary phosphate-binding elements for phosphoprotein enrichment.

Another enrichment strategy is the elimination of the labile phosphate group (β-elimination) under strongly basic conditions. Phosphate groups are replaced with biotin moieties, and proteins/peptides are then enriched on avidin supports. A major drawback of this approach is the inability to β-eliminate the phosphate group of tyrosine residues.

Kinases (ATPases) and GTPases can also be enriched, although this approach does not provide information on the proteins that these enzymes modify. These enrichment strategies use nucleotide derivatives that bind to the active site of kinases or GTPases (depending on the derivative used) and mediate the covalent attachment of a modified biotin (desthiobiotin). Because desthiobiotin exhibits reversible binding to streptavidin, labeled kinases or GTPases can be enriched from samples.

Specific enzymes can be enriched using consensus sequences of downstream proteins as probes. This approach is popular with enriching GTPases, in which the protein-binding domains of the downstream effectors of specific GTPases are fused to GST for the selective enrichment of distinct GTPases.

Quantitative Mass Spectrometry

Recent mass spectrometric strategies using stable isotope labeling by amino acids in cell culture (SILAC) or labeling of peptides in vitro with tandem mass tags have paved the way for the relative determination of changes in phosphorylation.

Absolute quantitation strategies are also available using isotopically "heavy" peptide standards. These approaches allow researchers to understand global phosphoproteomic changes in response to different stimuli or disease states. The diagram below illustrates the nature of complexity associated with protein sample preparation for MS. Phosphopeptide enrichment reduces sample complexity and is required prior to MS due to the low stoichiometry and poor ionization of phosphopeptides.

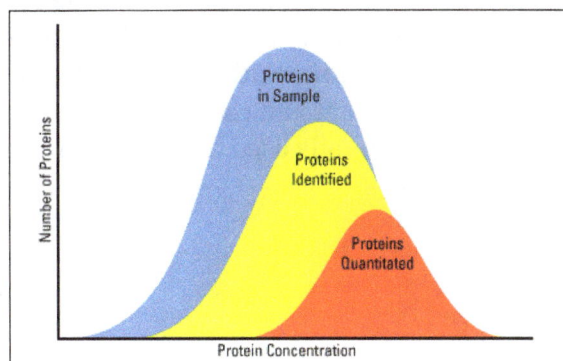

Protein availability for quantitative proteomic analysis is limited. Protein abundance and sample complexity are significant factors that affect the availability of proteins for mass spectrometric quantitation.

Phospho-Specific Antibody Development

A classical method of directly measuring protein phosphorylation involves the incubation of whole cells with radiolabeled ^{32}P-orthophosphate, the generation of cellular extracts, separation of proteins by SDS-PAGE, and exposure to film. This labor-intensive method requires many multi-hour incubations and the use of radioisotopes. Other traditional methods include 2-dimensional gel electrophoresis, a technique that assumes phosphorylation will alter the mobility and isoelectric point of the protein.

In light of these laborious methods, the development of phosphorylation-dependent antibodies was a welcome event for researchers. In 1981, the first documented phospho-antibody was produced in rabbits immunized with benzonyl phosphonate conjugated to keyhole limpet hemocyanin (KLH). This antibody broadly recognized proteins containing phosphotyrosine Ten years later, phosphorlyation state-specific (phospho-specific) antibodies were developed by immunizing rabbits with synthetic phosphopeptides representing the amino acid sequence surrounding the phosphorylation site of the target protein. The immune sera was applied to a peptide affinity column to generate a highly specific immunoreagent. The availability of phospho-specific antibodies has opened the door for the improvement of traditional methods as well as the development of new immunoassay techniques. The main caveat in utilizing phospho-specific antibodies in any technique is that successful detection is dependent on the specificity and affinity of the antibody for the phospho-protein of interest.

Western Blot

The Western blot is the most common method used for assessing the phosphorylation state of a protein, and most cell biology laboratories possess the equipment necessary to perform these experiments. Following separation of the biological sample with SDS- PAGE and subsequent transfer to a membrane (usually PVDF or nitrocellulose), a phospho-specific antibody can be used to identify the protein of interest for Western blot analysis). The typical Western blot protocol eliminates the hazards and waste disposal requirements associated with the use of radioisotopes. Many phospho-specific antibodies are quite sensitive and can readily detect the phosphorylated protein in a routine sample (e.g., 10-30 µg whole cell extract). Because the measured levels of a phospho-protein may change with treatment or through gel loading errors, researchers often utilize an antibody that detects the total level of the cognate protein (regardless of phosphorylation state) to determine the phosphorylated fraction relative to the total fraction and to serve as an internal loading control. Both chemiluminescent and colorimetric detection methods are common, and molecular weight markers are also generally used to provide information about protein mass.

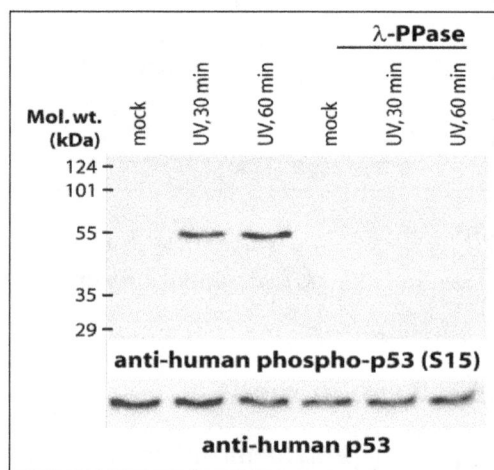

Phosphorylated p53 in CEM Cells. Human T lymphoblast CEM cells were exposed to UV-C light. Cellular extracts generated at 30 or 60 minutes post-irradiation were assessed by Western blot using rabbit anti-human phospho-p53 (S15) polyclonal antibody or goat anti-human p53 polyclonal antibody. Indicated samples were treated with lambda-phosphatase (lambda-PPase).

Enzyme-Linked Immunosorbent Assay (ELISA)

The ELISA has become a powerful method for measuring protein phosphorylation. ELISAs are more quantitative than Western blotting and show great utility in studies that modulate kinase activity and function. The format for this microplate-based assay typically utilizes a capture antibody specific for the desired protein, independent of the phosphorylation state. The target protein, either purified or as a component in a complex heterogeneous sample such as a cell lysate, is then bound to the antibody-coated plate. A detection antibody specific for the phosphorylation site to be analyzed is then added. These assays are typically designed using colorimetric or fluorometric detection. The intensity of the resulting signal is directly proportional to the concentration of phosphorylated protein present in the original sample. The phospho-specific ELISA technique confers several advantages over more traditional immunoblotting in the measurement of protein phosphorylation. First, results are easily quantifiable by utilizing a calibrated standard. Second, high specificity is possible due to the use of two antibodies specific for the target protein employed together in the sandwich format. Thirdly, the higher sensitivity often accomplished using ELISAs allows for smaller sample volumes and the detection of low abundance proteins. Finally, the microplate-based format allows for much higher throughput than traditional Western blotting. ELISAs generally provide an indirect measurement of kinase activity. However, variations in the technique described above use an immobilized capture antibody, substrate, and a phospho-substrate detection method for more direct measurements of kinase activity.

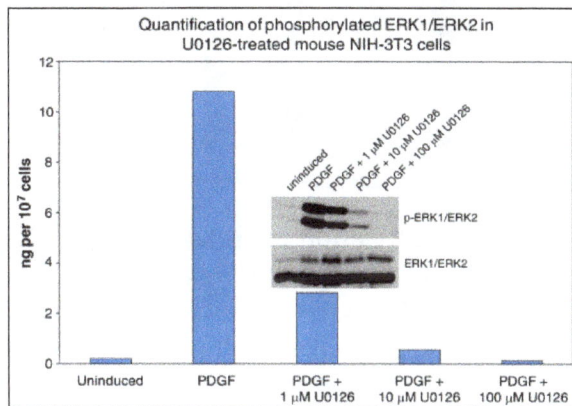

Phosphorylated ERK1/ERK2 in NIH-3T3 Cells. NIH-3T3 cells were treated with 100 ng/mL of human PDGF for 10 minutes in the presence or absence of the MEK1/2 inhibitor U0126. Following cell lysis, phosphorylated ERK1 and ERK2 were quantified with the Surveyor IC Immunoassay kit. The results are highly comparable between the Surveyor IC Immunoassay results and the amounts of phosphorylated ERK1/ERK2 relative to total ERK1/ERK2 detected by Western blot (inset).

Cell-based ELISA

Although in vitro biochemical kinase assays such as the typical sandwich ELISA are routinely used for hypothesis testing and drug screening, they cannot replicate the intracellular environment. Analyzing protein phosphorylation within intact cells may more accurately represent the status of specific signaling networks. Several immunoassays enabling the measurement of protein phosphorylation in the context of a whole cell have recently been developed. The cells are stimulated, fixed, and blocked in the same well. Phospho-specific antibodies are used to assess phosphorylation status using fluorometric or colorimetric detection systems. Furthermore, the phospho-protein and total protein are simultaneously detected in the same microplate well. Therefore, signals derived from the target protein can be normalized to that of the second protein, correcting for well-to-well variations and allowing phospho-protein levels to be accurately assessed and compared across multiple samples, similar to using phospho-specific and total protein antibodies in a traditional immunoblot. These assays bypass the need for the creation of cell lysates and are therefore more amenable to high throughput analyses.

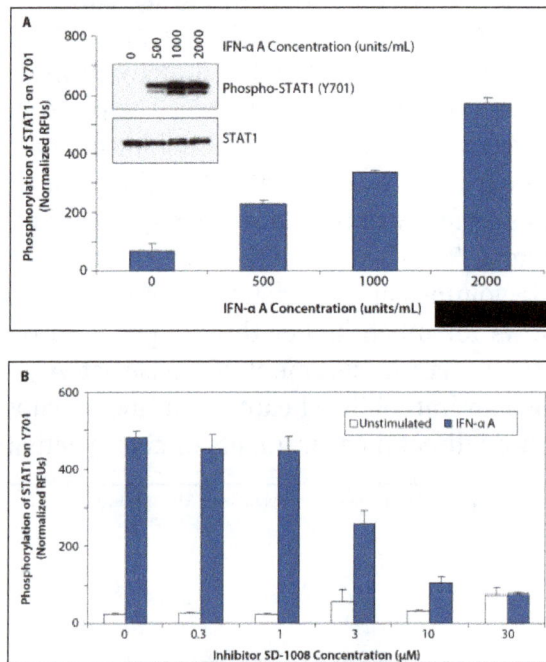

Measurement of STAT1 (Y701) Phosphorylation in HeLa Cells. (A) HeLa cells were treated with human IFN-alpha A for 20 minutes. After fixation of cells, phosphorylation of STAT1 (Y701) was determined and normalized to total STAT1 in the same well using the Phospho-STAT1 (Y701) Cell-Based ELISA kit. Values represent mean + range of duplicate determinations. Analysis of phosphorylated STAT1 and total STAT1 by Western blotting is also shown (inset A). (B) HeLa cells were pretreated for 30 minutes with the Janus kinase 2 (JAK2) inhibitor SD-1008 and then either left unstimulated or treated with 2000 units/mL IFN-alpha A for 20 minutes. Phosphorylation of STAT1 (Y701) was determined as in (A).

Intracellular Flow Cytometry and ICC/IHC

The traditional techniques of intracellular flow cytometry and immunocytochemistry/immunohistochemistry (ICC/IHC) are powerful tools for detecting phosphorylation events. Flow cytometry uses a laser to excite the fluorochrome used for antibody detection. Filter sets and fluorochromes with non-overlapping spectra must be carefully chosen when assessing multiple proteins in the same cell. Flow cytometry is advantageous because it allows for rapid, quantitative, single cell analysis. Proteins can be detected in a specific cell type within a heterogeneous population via cell surface marker phenotyping without the need to physically separate the cells. In this way, a small, rare population of cells may be analyzed without concern for cell loss or altered protein expression that may occur during a cell-sorting process for flow cytometry.

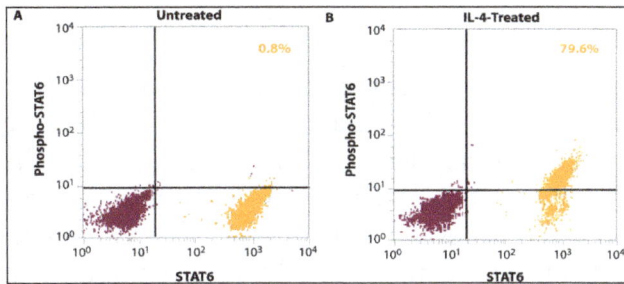

Detection of IL-4-induced STAT6 Phosphorylation by Intracellular Flow Cytometry. Human Daudi lymphoblastoid cells, (A) untreated or (B) activated with recombinant human IL-4 were fixed and permeabilized with methanol. Simultaneous detection (orange) of total (x axis) and phosphorylated (y axis) STAT6 was performed by co-staining cells with allophycocyanin-conjugated anti-STAT6 and phycoerythrin-conjugated anti-phospho-STAT6, respectively. The percent of cells positive for phospho-STAT6 is indicated in each treatment (upper right). Staining with isotype controls highlights the specificity of the STAT6 antibodies.

ICC generally refers to protein detection by microscopy in cultured cells, while IHC refers to protein detection in intact tissue sections. Like flow cytometry, these techniques allow for the assessment of multiple proteins within a cell or tissue provided that adequate attention is given to avoid overlapping fluorescence spectra or color. Both fluorescent and colorimetric detection techniques are commonly used. In contrast to other formats for monitoring phosphorylation, ICC is usually the method of choice for determining intracellular localization for ICC/IHC). Both flow cytometry and ICC/IHC require high-affinity and high-specificity antibodies, blocking steps, controls, and antibody titration to eliminate ambiguous results due to non-specific binding.

Detection of phospho-proteins by flow cytometry and ICC require that the protein is stable and accessible to the antibody. Cells are usually stimulated and fixed with formaldehyde or paraformaldehyde to cross-link the phospho-proteins and stabilize them for analysis. The fixed cells must then be permeabilized to allow for entry of phospho-specific antibodies into the cells. Different permeabilization techniques are often useful for various subcellular locations. A mild detergent will allow for detection of cytoplasmic proteins, while alcohol may be required for antibody access to nuclear proteins. Alcohol permeabilization may also enhance phospho-protein detection using peptide specific antibodies due to the denaturing property of alcohol.

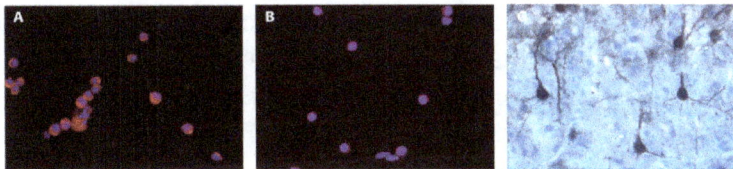

Detection of Phosphorylated Proteins Using ICC/IHC. Human Daudi lymphoblastoid cells were treated with (A) recombinant human IL-4 or (B) left untreated. Phosphorylated STAT6 was detected using anti-human phospho-STAT6 polyclonal antibody, followed by staining with NorthernLights™ 557-conjugated goat anti-rabbit IgG and DAPI nuclear staining (blue). (C) Phosphorylated ERK1/ERK2 was detected in a section of inflamed rat brain cortex using anti-human/mouse/rat phospho-ERK1/ERK2 polyclonal antibody. The tissue was stained with the anti-rabbit HRP-DAB Cell and Tissue Staining Kit and counterstained with haematoxylin (blue).

Multi-Analyte Profiling

Mass spectrometric techniques such as collision-induced dissociation (CID) and electron transfer dissociation (ETD) provide comprehensive parallel analysis of peptide sequences and post-translational modifications such as phosphorylation. These techniques are labor-intensive, and strategies for comprehensive phosphorylation analysis may not be needed if particular pathways are of primary interest. This has led to the development of several novel methods for measuring protein phosphorylation of multiple analytes simultaneously. In general, these involve the use of phospho-specific antibodies and include microplate-based and membrane-based detection formats. The obvious benefit of these assays is that throughput capability is greatly enhanced by bypassing the need for running multiple individual Western blots or traditional ELISA-based assays. These techniques are also known for providing more data while requiring very little sample volume. In trade, protein profiling assays are typically recognized as being less sensitive than their more conventional counterparts due to potential antibody cross-reactivity.

Glycosylation

Glycosylation refers to the attachment of sugar moieties to proteins and is a post-translational modification (PTM) that provides greater proteomic diversity than other PTMs. Glycosylation is critical for a wide range of biological processes, including cell attachment to the extracellular matrix and protein–ligand interactions in the cell. This PTM is characterized by various glycosidic linkages, including N-, O- and C-linked glycosylation, glypiation (GPI anchor attachment), and phosphoglycosylation. Glycoproteins can be detected, purified and analyzed by different strategies, including glycan staining and visualization, glycan crosslinking to agarose or magnetic resin for labeling or purification, or proteomic analysis by mass spectrometry, respectively.

Glycosylation is a critical function of the biosynthetic-secretory pathway in the endoplasmic reticulum (ER) and Golgi apparatus. Approximately half of all proteins typically expressed in a cell undergo this modification, which entails the covalent addition of sugar moieties to specific amino acids. Most soluble and membrane-bound proteins expressed in the endoplasmic reticulum are glycosylated to some extent, including secreted proteins, surface receptors and ligands, and organelle-resident proteins. Additionally, some proteins that are trafficked from the Golgi to the cytoplasm are also glycosylated. Lipids and proteoglycans can also be glycosylated, significantly increasing the number of substrates for this type of modification.

Scope

Protein glycosylation has multiple functions in the cell. In the ER, glycosylation is used to monitor the status of protein folding, acting as a quality control mechanism to ensure that only properly folded proteins are trafficked to the Golgi. Sugar moieties on soluble proteins can be bound by specific receptors in the trans Golgi network to facilitate their delivery to the correct destination. These sugars can also act as ligands for receptors on the cell surface to mediate cell attachment or stimulate signal transduction pathways. Because they can be very large and bulky, oligosaccharides can affect protein–protein interactions by either facilitating or preventing proteins from binding to cognate interaction domains. Because they are hydrophilic, they can also alter the solubility of a protein.

Distribution

Glycosylated proteins (glycoproteins) are found in almost all living organisms that have been studied, including eukaryotes, eubacteria and archae. Eukaryotes have the greatest range of organisms that express glycoproteins, from single-celled to complex multicellular organisms.

Glycoprotein Diversity

Glycosylation increases the diversity of the proteome to a level unmatched by any other post-translational modification. The cell is able to facilitate this diversity, because almost every aspect of glycosylation can be modified, including:

- Glycosidic linkage—the site of glycan (oligosaccharide) binding.

- Glycan composition—the types of sugars that are linked to a particular protein.

- Glycan structure—branched or unbranched chains.

- Glycan length—short- or long-chain oligosaccharides.

Glycosylation is thought to be the most complex post-translational modification because of the large number of enzymatic steps involved. The molecular events of glycosylation include linking monosaccharides together, transferring sugars from one substrate to another and trimming sugars from the glycan structure. Unlike other cell processes such as transcription or translation, glycosylation is non-templated, and thus, all of these steps do not necessarily occur during every glycosylation event. Instead of using templates, cells rely on a host of enzymes that add or remove sugars from one molecule to another to generate the diverse glycoproteins seen in a given cell. While it may seem chaotic because of all of the enzymes involved, the different mechanisms of glycosylation are highly-ordered, step-wise reactions in which individual enzyme activity is dependent upon the completion of the previous enzymatic reaction. Because enzyme activity varies by cell type and intracellular compartment, cells can synthesize glycoproteins that differ from other cells in glycan structure.

Enzymes that transfer mono- or oligosaccharides from donor molecules to growing oligosaccharide chains or proteins are called glycosyltransferases (Gtfs). Each Gtf has specificity for linking a particular sugar from a donor (sugar nucleotide or dolichol) to a substrate and acts independent of other Gtfs. These enzymes are broad in scope, as glycosidic bonds have been detected on almost every protein functional group, and glycosylation has been shown to incorporate most of the commonly occurring monosaccharides to some extent.

Glycosidases catalyze the hydrolysis of glycosidic bonds to remove sugars from proteins. These enzymes are critical for glycan processing in the ER and Golgi, and each enzyme shows specificity for removing a particular sugar (e.g., mannosidase).

Types of Glycosylation

Glycopeptide bonds can be categorized into specific groups based on the nature of the sugar–peptide bond and the oligosaccharide attached, including N-, O- and C-linked glycosylation, glypiation and phosphoglycosylation. Because N- and O-glycosylation and glypiation are the most commonly detected types of glycosylation.

Types of Glycosylation	
N-linked	Glycan binds to the amino group of asparagine in the ER.
O-linked	Monosaccharides bind to the hydroxyl group of serine or threonine in the ER, Golgi, cystosol and nucleus.
Glypiation	Glycan core links a phospholipid and a protein.
C-linked	Mannose binds to the indole ring of tryptophan.
Phosphoglycosylation	Glycan binds to serine via phosphodiester bond.

Proteins are not restricted to a particular type of glycosylation. Indeed, proteins are often glycosylated at multiple sites with different glycosidic linkages, which depends on multiple factors including those described below.

1. Enzyme availability: Glycosylation is controlled by moving proteins to areas with different enzyme concentrations; the cell sequesters enzymes into specific compartments to regulate their activity. For example, after a protein is N-glycosylated in the ER, glycan processing occurs in a stepwise fashion by trafficking proteins to distinct Golgi cisternae that contain high concentrations of specific Gtfs and glycosidases.

2. Amino acid sequence: Besides the requirement for the right amino acid (e.g., Asn for N-linked; Ser/Thr for O-linked), many enzymes have consensus sequences or motifs that enable formation of the glycosidic bond.

3. Protein conformation (availability): As proteins are synthesized, they begin to fold into their nascent secondary structure, which can make specific amino acids inaccessible for glycosidic binding. Thus, the target amino acids must be conformationally accessible for glycosylation to occur.

Methods to Detect and Analyze Glycoproteins

The influence that glycoproteins have on biological processes and disease states continues to expand, spurring the development of detection and analytical strategies with increasing sensitivity and throughput to better understanding their diverse structures and biochemistry. Because glycoproteins are a combination of a protein and oligosaccharides, they are more complex to analyze than non-glycosylated proteins. Additionally, the vast diversity in glycan structure and composition add an additional level of complexity to glycoprotein analysis.

Glycan Staining or Labeling

Because of their structure, glycan sugar moieties are not reactive to staining or labeling molecules. To overcome this problem, sugar groups can be chemically restructured with periodic acid, which oxidizes vicinal hydroxyls on sugars (especially sialic acid) to aldehydes or ketones that are then reactive to multiple dyes. The periodic acid-Schiff (PAS) stain uses this reaction to detect and quantify glycoproteins in various biological samples. Periodic acid can also be used to make sugars reactive towards crosslinkers, which can be covalently bound to labeling molecules (e.g., biotin) or immobilized support (e.g., streptavidin) for detection or purification. The two representative protein gels below compare the use of glycan to Coomassie protein staining.

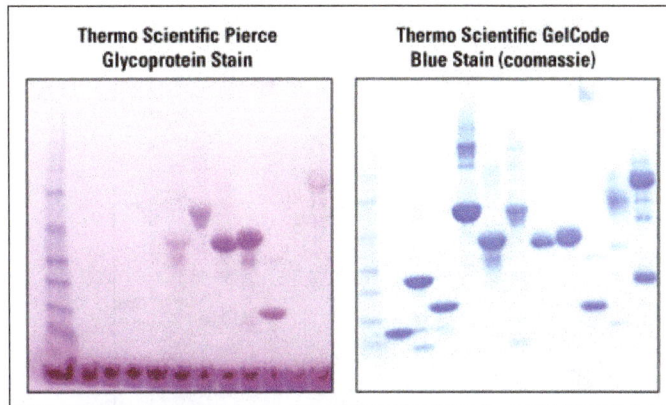

Sensitive, specific staining of glycosylated proteins with the Glycoprotein Staining Kit.

Glycoprotein Purification or Enrichment

Lectins can be used to detect and analyze glycoprotein function. These glycan-binding proteins have high specificity for distinct sugar moieties. As described previously, lectins facilitate protein folding in the ER, but they are also critical for cell–cell and pathogen–cell attachment. While anti-glycan antibodies can also bind sugar moieties, lectins are used more often because they are less expensive, better characterized and more stable than antibodies. Like antibodies, lectins can be conjugated to probes such as horseradish peroxidase, fluorophores and biotin and immobilized to solid support including streptavidin and Thermo Scientific NeutrAvidin Protein. Some of the common uses of lectins include:

- Glycoprotein identification.

- Glycoprotein purification/enrichment.

- Characterization of cell surface glycoconjugates (e.g., glycoproteins, glycolipids, GPI-anchored molecules):

  o Detect relative abundance.

  o Identify tissue and cellular localization.

- Glycosylation mutant generation.

- Gtf and glycosidase activity analysis.

Many lectins are commercially available for use in these applications. The most popular lectin is concanavalin A (ConA) from the Jackbean, but other lectins, including Jacalin, wheat germ agglutinin (WGA) and lentil lectin (LCA), are also widely available in different commercial kits. These representative silver stained gels compare results generated using different glycoprotein isolation kits.

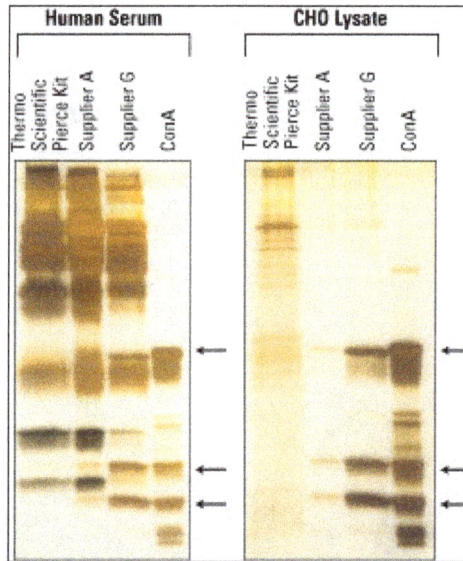

Glycoprotein isolation from human serum and cell lysate—performance comparison of kits using ConA resin.

Glycoproteome and Glycome Analysis by Mass Spectrometry

Glycoproteins are unique from other post-translationally modified proteins because they are a combination of a protein portion and a glycan portion, both of which can comprise a significant proportion of the molecular weight of the molecule. Thus, glycoproteins can be analyzed as a whole or as individual components. Glycoproteomics is the global analysis of glycosylated proteins and integrates glycoprotein enrichment and proteomic analysis for the systematic identification and quantitation of glycoproteins in complex systems. This subset of proteomics differs from glycomics, which is restricted to all glycans in a system (i.e., the glycome).

As with other proteomic analyses, glycoprotein identification and quantitation is performed using mass spectrometry. The basic pipeline for glycoproteomic analysis includes:

- Glycoprotein or glycopeptides enrichment.

- Multidimensional separation by liquid chromatography (LC).

- Tandem mass spectrometry.

- Data analysis via bioinformatics.

This approach can be performed before or after enzymatic cleavage of glycans via endoglycanase H (endo H) or peptide-N4-(N-acetyl-beta-glucosaminyl) asparagine amidase (PNGase), depending on the type of experiment. Quantitative comparative glycoproteome analysis can be performed by differential labeling with stable isotope labeling by amino acids in cell culture (SILAC) reagents. Additionally, absolute quantitation by selected reaction monitoring (SRM) can be performed on targeted glycoproteins using isotopically labeled, "heavy" reference peptides.

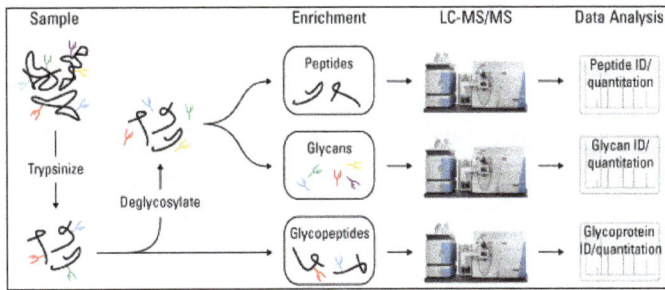

Glycoproteomic analysis. Glycoproteins are first digested into glycopeptides and either analyzed directly by liquid chromatography and tandem mass spectrometry (LC-MS/MS) or first deglycosylated and then enriched for glycans or peptides prior to analysis.

N-glycosylation

There are many types and mechanisms of glycosylation, and the most common type is N-glycosylation. Glycosylation is often characterized as a post-translational modification. While this is true with other types of glycosylation, N-glycosylation often occurs co-translationally, in that the glycan is attached to the nascent protein as it is being translated and transported into the ER. The "N" in the name of this type of glycosylation denotes that the glycans are covalently bound to the carboxamido nitrogen on asparagine (Asn or N) residues.

Because the ER is the site of translation and processing of most membrane-bound and secreted proteins, it is not surprising that most of these are N-linked glycoproteins. Besides being the most common type of glycosylation (90% of glycoproteins are N-glycosylated), N-linked glycoproteins also have large and often extensively branched glycans that undergo multiple processing steps after being bound to proteins.

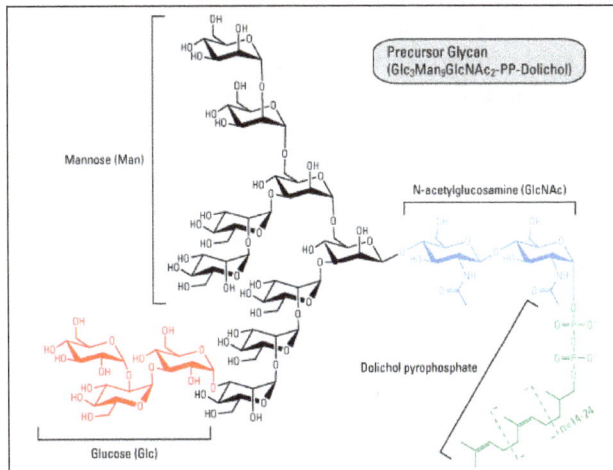

The structure of N-glycosylation.

N-glycosylation is conserved across eukaryotes and archae, and a considerable number of the enzymes and processes involved are also conserved across the different species. N-glycosylation can be broken down into separate events, as follows:

- Precursor glycan assembly,

- Attachment,

- Trimming,

- Maturation.

Different enzymes are required for each step in during glycosylation, which facilitate diversity in the glycans that are generated. But N-glycosylation initially occurs identically for all proteins, and the diversity does not manifest until the subsequent trimming and glycan maturation.

Precursor Glycan Assembly

Oligosaccharides attached via N-glycosidic linkages are derived from a 14-sugar precursor molecule comprised of N-acetylglucosamine (GlcNAc), mannose (Man) and glucose (Glc). These sugars are added consecutively onto dolichol, a polyisoprenoid lipid carrier embedded in the ER membrane. The first 7 sugars are donated from sugar nucleotides (UDP- and GDP-sugars) in the cytoplasm and bound to dolichol via a pyrophosphate linkage (-PP-). After the Man5GlcNAc2-PP-dolichol intermediate is completed, the entire complex is flipped into the lumen of the ER, after which the final 7 sugars are donated from Man- and Glc-P-dolichol molecules to make the Gcl3Man9GlcNAc2-PP-dolichol precursor glycan.

Glycan Attachment

Glycosylation is often characterized as a post-translational modification. While this is true with other types of glycosylation, N-glycosylation often occurs co-translationally, in that the glycan is attached to the nascent protein as it is being translated and transported into the ER. The "N" in the name of this type of glycosylation denotes that the glycans are covalently bound to the carboxamido nitrogen on asparagine (Asn or N) residues.

Because the ER is the site of translation and processing of most membrane-bound and secreted proteins, it is not surprising that most of these are N-linked glycoproteins. Besides being the most common type of glycosylation (90% of glycoproteins are N-glycosylated), N-linked glycoproteins also have large and often extensively branched glycans that undergo multiple processing steps after being bound to proteins.

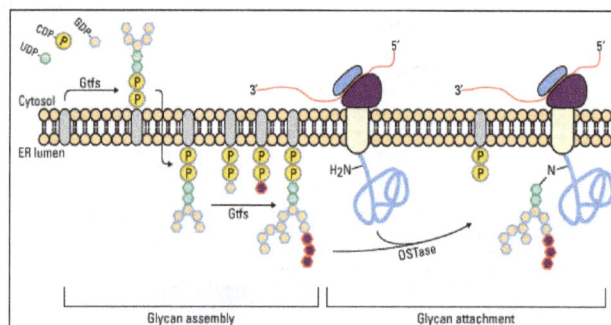

Glycan assembly and attachment. Precursor glycan synthesis begins on the cytosolic face of the endoplasmic reticulum (ER) and is completed after the structure is flipped into the ER lumen. Oligosaccharide transferase (OSTase) then transfers the precursor glycan to the Asn residue on the nascent protein.

One aspect to note is that not all Asn residues with the predicted consensus sequence are glycosylated. N- to C-terminal protein synthesis results in transport of the growing polypeptide into the ER in the same orientation, and protein folding occurs soon after the polypeptide enters the ER.

Therefore, as protein folding increases, OSTase is less able to access the consensus sequence for glycan transfer. Indeed, more N-terminal Asn residues are glycosylated than C-terminal Asn residues.

Glycan Trimming in the ER

Oligosaccharides are trimmed in both the ER and Golgi by glycosidases via hydrolysis. Glycan trimming in the ER, though, serves a different purpose than trimming in the Golgi.

In the ER, sugar hydrolysis is used to both monitor protein folding and indicate when proteins should be degraded. Glucosidases I and II remove 2 terminal Glc from the precursor glycan, after which calnexin and calreticulin, which are membrane-bound and soluble (respectively) sugar-binding lectins, bind to the nascent glycoprotein via the remaining Glc and act as chaperones to help the protein fold properly. The final Glc is soon hydrolyzed by glucosidase II, releasing the glycoprotein from the chaperone. Non-native–folded proteins are recognized by UDP-glucose glycoprotein glucosyltransferase, which transfers a Glc to the glycoprotein, and the protein again is bound to the lectin chaperones to facilitate proper protein folding. This cycle of Glc addition and removal continues until the protein is correctly folded, at which time it is not reglycosylated, and the glycoprotein is trafficked to the Golgi for further processing. The glycan structure for all properly folded glycoproteins that proceed to the Golgi is Man9GlcNAc2 in higher eukaryotes.

An ER-resident mannosidase (ERManI) plays a key role in identifying proteins that are unable to fold properly. Proteins that lose 3–4 mannose residues in the ER via ERManI activity are transported out of the ER and deglycosylated by glycanase N (removes the entire glycan en bloc) and delivered to ER-associated degradation (ERAD). It is thought that ERManI acts as a timer of sorts, because it has a slow rate of mannose hydrolysis that allows nascent proteins multiple rounds of reglycosylation to attempt to fold properly before mannose residues are removed and the protein is targeted for degradation.

Glycan Maturation in the Golgi

To this point during glycosylation, all N-linked glycoproteins have the same precursor glycan structure. Glycan processing in the Golgi apparatus combines both trimming and adding sugars to diversify the glycans on individual glycoproteins. As with precursor glycan biosynthesis, this maturation pathway to generate diverse oligosaccharides is highly ordered, such that each step is dependent upon the previous step. To this end, the Golgi segregates specific enzymes into different cisternae to facilitate this step-wise process.

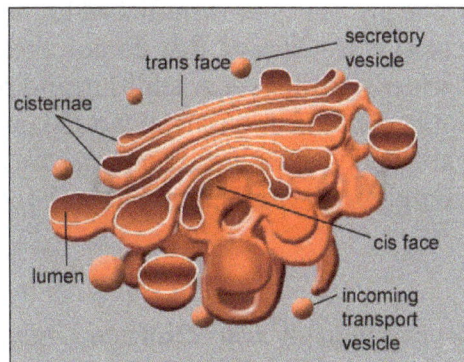

Golgi enzyme compartmentalization. Enzymes that mediate glycan processing in the Golgi apparatus are segregated into distinct cisternae to ensure that glycosylation occurs in a step-wise fashion.

The final glycan structures can be broadly separated into two groups:

- Complex oligosaccharides—contain multiple sugar types.

- High-mannose oligosaccharides—multiple mannose residues.

- Hybrid—branches of both high mannose and complex oligosaccharides.

Glycans destined to be complex oligosaccharides are trimmed by Golgi mannosidase I and II and glycosylated by GlcNAc transferase, resulting in a common core region. The core then becomes the substrate for multiple Gtfs that consecutively transfer sugar moieties from sugar nucleotides to build variable-length and -branched oligosaccharide chains of GlcNAc, galactose (Gal), N-acetyl-neuraminic acid (NANA or sialic acid) and fucose. Any glycoproteins that progress through this processing from the common core stage become resistant to glycan removal by endoglycosidase H (endo H), which is used experimentally to determine if glycoproteins contain high-mannose or complex oligosaccharides.

Unlike complex oligosaccharides, high-mannose oligosaccharides do not carry other sugar moieties, although some of the Man residues are often trimmed by Golgi mannosidase I. Whether a glycan is processed into a complex oligosaccharide rather than remaining a high-mannose oligosaccharide is dependent upon the accessibility of the processing enzymes to the glycan, which can be hindered by the glycoprotein conformation. Some glycoproteins have hybrid oligosaccharides, comprising a combination of complex and high-mannose glycans.

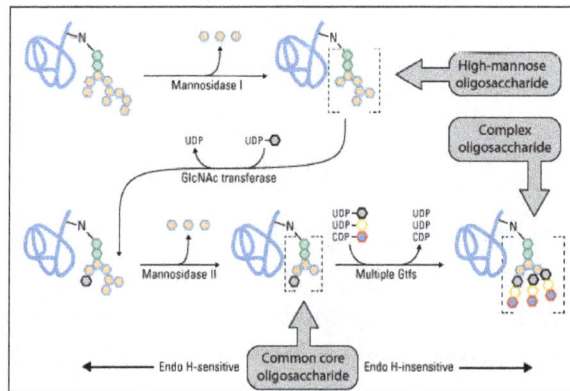

Glycan maturation.

After initial trimming in the ER, the glycoprotein is trafficked to the Golgi, where Golgi mannosidase I removes multiple mannose sugars. Glycans that do not undergo further glycosylation are called high-mannose oligosaccharides. Further sugar addition and removal yields a common core oligosaccharide onto which multiple Gtfs add different sugars to generate the highly variable complex oligosaccharides. Glycan maturation beyond the common core provides endo H insensitivity. Glycans can be high-mannose, complex or a combination of both (i.e., hybrid oligosaccharide).

O-glycosylation

While N-glycosylation is the most common glycosidic linkage, O-glycoproteins also play a key role in cell biology. This type of glycosylation is essential in the biosynthesis of mucins, a family of heavily O-glycosylated, high-molecular weight proteins that form mucus secretions. O-glycosylation is also critical for the formation of proteoglycan core proteins that are used to make extracellular matrix components. Additionally, antibodies are often heavily O-glycosylated.

O-glycosylation occurs post-translationally on serine and threonine side chains in the Golgi apparatus. N-glycosylation does not preclude the other from occurring, as O-glycosylation commonly occurs on glycoproteins that were N-glycosylated in the ER. Besides the different linkage, O-glycosylation also differs in the method of glycosylation. While a precursor glycan is transferred en bloc to Asn via N-glycosylation, sugars are added one-at-a-time to serine or threonine residues. O-glycosylation can also occur on hydroxylysine and hydroxyproline, oxidized forms of lysine and proline, respectively, which are found in collagen. Additionally, O-linked glycans usually have much simpler oligosaccharide structures than N-linked glycans.

Mechanism

The O-glycosidic mechanism is not as complex as that of N-glycosylation. Proteins trafficked into the Golgi are most often O-glycosylated by N-acetylgalactosamine (GalNAc) transferase, which transfers a single GalNAc residue to the β-OH group of serine or threonine. To date, there is no known consensus sequence for this enzyme, although structural motifs have been characterized. Some proteins are O-glycosylated with GlcNAc, fucose, xylose, galactose or mannose, depending on the cell and species. As with N-glycosylation, sugar nucleotides are used as monosaccharide donors for O-glycosylation. Following this first sugar, a highly variable number of sugars (from only a few to greater than 10) are consecutively added to the growing glycan chain. O-glycosylation can also occur in the cytosol and nucleus to regulate gene expression or signal transduction through other Gtfs.

Glypiation

The covalent attachment of a glycosylphosphatidylinositol (GPI) anchor is a common post-translational modification that localizes proteins to cell membranes. This special kind of glycosylation is widely detected on surface glycoproteins in eukaryotes and some archae.

GPI anchors consist of a:

- Phosphoethanolamine linker that binds to the C-terminus of target proteins.

- Glycan core structure.

- Phospholipid tail that anchors the structure in membrane.

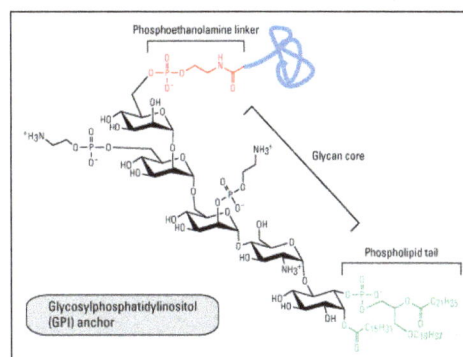

The structure of glypiation.

Both the lipid moiety of the tail and the sugar residues in the glycan core have considerable variation, demonstrating vast functional diversity that includes signal transduction, cell adhesion and immune recognition. GPI anchors can also be cleaved by enzymes such as phospholipase C to regulate the localization of proteins that are anchored at the plasma membrane.

Mechanism

Similar to the precursor glycan used for N-glycosylation, GPI anchor biosynthesis begins on the cytoplasmic leaflet of the ER and is completed on the luminal side. During this process, 3–4 Man and various other sugars (e.g., GlcNAc, Gal) are built onto a phosphatidylinositol (PI) molecule embedded in the membrane using sugars donated from sugar nucleotides and dolichol-P-mannose outside and inside the ER, respectively. Additionally, 2–3 phosphoethanolamine (EtN-P) linker residues are donated from phosphatidylethanolamine in the ER lumen to facilitate binding of the anchor to proteins.

Proteins destined to be glypiated have 2 signal sequences:

- An N-terminal signal sequence that directs co-translational transport into the ER.

- A C-terminal signal sequence that is recognized by a GPI transamidase (GPIT).

GPIT does not have a consensus sequence but instead recognizes a C-terminal sequence motif that enables it to covalently attach a GPI anchor to an amino acid in the sequence. This C-terminal sequence is embedded in the ER membrane immediately after translation, and the protein is then cleaved from the sequence and attached to a preformed GPI anchor.

C-glycosylation

C-mannosylation represents a different approach to glycosylation, because the reaction forms carbon–carbon bonds rather than carbon-nitrogen or carbon-oxygen bonds. C-mannosyltransferase (c-Mtf) links C1 of mannose to C2 of the indole ring of tryptophan. The enzyme recognizes the specific sequence Trp-X-X-Trp and transfers a mannose residue from dolichol-P-Man to the first Trp in the sequence.

C-mannosylation has been detected in multiple cell lines and rat liver microsomes. Specific proteins that are C-glycosylated include Trp2 in RNAse, the erythropoietin receptor and IL-12B. The biological function of C-glycosylation is unknown, but current research focuses on the synthesis of C-glycosylated molecules by plants, insects and bacteria for drug discovery, because they are resistant to metabolic hydrolysis.

Phosphoglycosylation

This type of post-translational modification is limited to parasites (e.g., Leishmania and Trypanosoma) and slime molds (e.g., Dictyostelium) and is characterized by the linking of glycans to serine or threonine via phosphodiester bonds. In some parasitic species, such as Leishmania, phosphoglycosylation is the most abundant post-translational modification and is used to make proteophosphoglycans (PPGs), which are critical for protection against host complement and promote parasite aggregation in the host. Similar to N-glycosylation, phosphoglycosylation occurs by transfer of a prefabricated phosphoglycan from a membrane-bound molecule via a phosphoglyco-syltransferase (PTase), although the exact structure and enzyme varies by species.

Post-Glycosylation Modifications

Besides multiple types of glycosylation occurring on the same protein, glycans can be further modified to increase the diversity of glycoproteins in a given proteome. These modifications include:

- Sulfation at Man and GlcNAc residues in the production of glycosaminoglycans (GAGs), which are components of proteoglycans in the extracellular matrix.

- Acetylation of sialic acid to facilitate protein-protein interactions.

- Phosphorylation, such as with Man residues on precursor lysosomal proteins (mannose 6-phosphate) to ensure trafficking to lysosomes by binding to mannose 6-phosphate receptor (M6PR) in the Golgi.

S-Nitrosylation

S-Nitrosylation refers to a covalent attachment of an NO moiety to sulfhydryl residues of proteins. The sulfhydryl residues belongs to a subset of specific cysteine residues in proteins, the resulting SNO is an S-nitrosoprotein. SNOs have a short half-life in the cytoplasm because of the host of reducing enzymes, including glutathione (GSH) and thioredoxin, that denitrosylate proteins. Therefore, SNOs are often stored in membranes, vesicles, the interstitial space and lipophilic protein folds to protect them from denitrosylation. For example, caspases, which mediate apoptosis, are stored in the mitochondrial intermembrane space as SNOs.

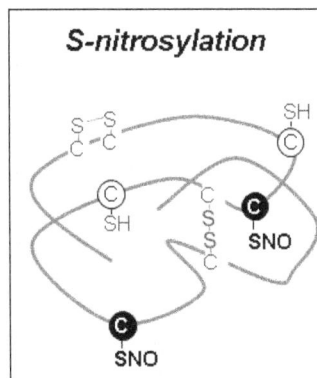

Because proteins may contain multiple cysteines and due to the labile nature of SNOs,
S-nitrosylated cysteines can be difficult to detect and distinguish from non-S-nitrosylated amino acids.

Similar to phosphorylation, S-Nitrosylation is a reversible process. The denitrosylation is an enzymatic catalyzing process that reverses the S-Nitrosylation process. However, S-nitrosylation is not a random event, and only specific cysteine residues are S-nitrosylated. Under physiologic conditions, protein S>-nitrosylation and SNOs provide protection preventing further cellular oxidative and nitrosative stress. Aberrant S-Nitrosylation may lead to protein misfolding, synaptic damage, and apoptosis. Dysfunction of the SNO signaling has been implicated in the pathogenesis of many diseases, such as Alzheimer's disease cardiovascular diseases.

Methylation

Proteins can be altered by a diverse set of post-translational modifications. These include the methylation of arginine residues, the methylation, acetylation, ubiquitylation or sumoylation of lysine residues, or prolyl-hydroxylation, and many others. In addition, the presence of multiple post-translational modifications on one protein can have combinatorial effects which can fine tune the regulation of intra-molecular interactions of domain containing proteins.

Today, protein methylation is of wide interest to the scientific field. This hasn't been always so. Researchers began to study protein methylation in the 1960s, and by the early 1980s, it was known that lysine, arginine, histidine and dicarboxylic amino acids were post-translationally modified. Highly specific enzymes called methyltransferases are known to be responsible for the selective transfer of a methyl group to a targeted molecule. With the availability of modern molecular biology techniques starting in the mid 1990s, is has now become clear that protein methylation is involved in many important functions, including gene regulation and signal transduction.

In chemistry, biochemistry, and biology methylation refers to the addition of methyl groups ($-CH_3$; the addition of 12 mass units or delta mass = 15 dalton) to organic compounds. This reaction type is a specific case of alkylation. The term alkylation refers to the transfer of an alkyl group, such as the isopropyl group, $-CH(CH_3)_2$, or the methyl group, $-CH_3$, from one molecule to another. Alkyl groups may be transferred as alkyl groups with a positively charged carbon atom, a free radical, a carbanion or a carbine. In chemistry, some typical methylation reagents are dimethylsulfate, methanol, methyl halides and diazomethane.

However, it is well established now that many different molecules present in a cell can be methylated. For example, scientists now know that methylation of DNA is epigenetically inherited. The methylation of DNA turned out to be an important regulator of gene transcription. The addition of methyl groups to DNA typically occurs at CpG islands or regions. In addition, methylation can also occur in RNA molecules, for example at conserved sequence regions in 18S rRNA. Methylation of cytosine residues in DNA to form 5-methylcytosine has widespread effects on gene expression due to recruiting specific DNA-binding proteins.

In cellular proteins many protein or peptide motifs that contain lysine residues can be methylated or acetylated. These modified motifs can lead to the recognition by other protein domains such as chromo-domains or bromo-domains. Such domains are found in proteins regulating chromatin structure and gene expression. For example, the flexible N-terminal and C-terminal ends of histones are known to contain lysine modifications important for the coupling of histones to changes in chromatin organization and the epigenetic control of gene expression. Typically, a single chromo- or bromo-domain recognizes a suitable modified lysine residue within a short peptide motif sequence.

The monomethylation of a lysine side chain is illustrated in figure and the structure of S-Adenosyl methionine (SAM), the primary methyl donor molecule in cellular metabolism for numerous biochemical reactions is shown in figure below.

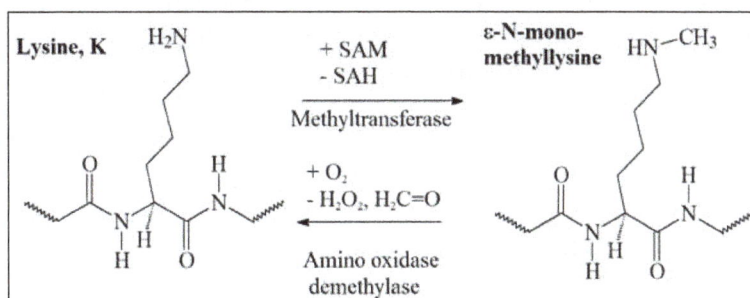

Monomethylation of a lysine on its ε-amino group. Adenosyl methionine (SAM), a molecule composed of adenosine and methionine, is the primary methyl group donor in metabolism. Donation of the methyl group, in this reaction to the ε-amino group of a lysine, transforms SAM into S-adenyl homocysteine (SAH).

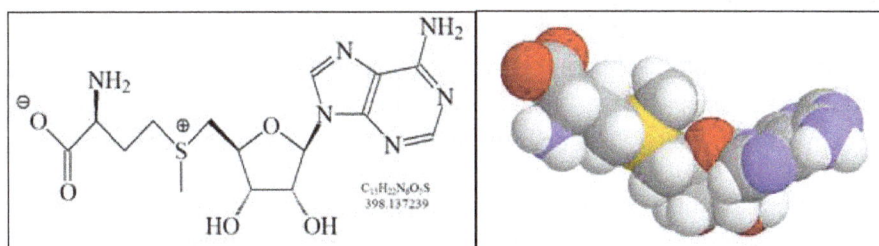

S-Adenosyl methionine (SAM), the primary methyl donor molecule in cellular metabolism for numerous biochemical reactions. During the methionine cycle, methionine is converted to SAM. After transfering its methyl group to a target molecule, SAM is converted to S-adenosyl homocysteine (SAH), which is then further converted to homocysteine. Homocysteine is either converted back to methionine, or enters the trans-sulfuration pathway to form other sulfur-containing amino acids.

The study of lysine methylation of histones has been quite fruitful in recent years after histone lysine methyltransferases were discovered. Histone lysine methylation, either activate or repress gene expression depending on the status of the methylated lysine and its position. However, some of the recently discovered lysine methyl transferases target not only histones but other proteins as well. For example, Set9, a SET domain-containing lysine methyltransferase, was initially found to target histone H3 lysine 4 for mono-methylation but was subsequently shown to target a variety of non-histone proteins as well. Some of its targets are transcription factors.

The SET domain is a 130 to 140 amino acid, evolutionary well conserved sequence motif initially characterized in the Drosophila proteins Su(var)3-9, Enhancer-of-zeste and Trithorax, from which the acronym SET is derived. In addition, the SET domain is found in proteins of diverse functions ranging from yeast to mammals, but also in some bacteria and viruses. Several structures of SET domain proteins have been reported over the past year. SET domains are folded in a novel way, and adjacent domains are used for both structural stabilization and the completion of their active sites. In addition, the cofactor S-adenosyl-L-methionine and peptide substrates bind on opposite faces of the SET domain. Furthermore, the side chain of the target lysine approaches the transferred methyl group through a narrow channel that passes through the middle of the domain.

Acetylation

Acetylation is a vital chemical reaction that is important for co-translational and post-translational modification of proteins. Once the proteins are formed in their rudimentary forms of long poly-peptide chains, they undergo several chemical reactions to form the final three dimensional structures of proteins. Acetylation is one such reaction. Some modifications include those for histones, p53, and tubulins.

N-alpha-terminal Acetylation

This is the acetylation reaction of the N-terminal alpha-amine of proteins. This is a common reaction seen in eukaryotes. Over half (40 to 50 percent) of yeast proteins and nearly all (80 to 90 percent) of human proteins are modified in this manner. This reaction has been conserved throughout evolution and has not changed much.

The reactions are mediated by N-alpha-acetyltransferases (NATs), a sub-family of the GNAT superfamily of acetyltransferases. This superfamily includes histone acetyl transferases. These NATs transfer the acetyl group from acetyl-coenzyme A to the amine group.

There are three types of N-acetyletransferases. These are labelled A, B and C. These have been extensively studied in yeast. Each subtype is specific for its substrates. These NATs are associated with the ribosome, where they acetylate the newly formed and unmodified polypeptide chain. Proteins such as actin and tropomyosin are especially dependent of NAT B acetylation to form proper actin filaments.

Humans also have the NAT A and NAT B complexes. NAT A complex activities have been associated with hypoxia-response and beta-catenin pathway that have been linked to cancer pathologies. NATA has been found to be over-expressed in papillary thyroid cancers and neuroblastomas. The human NAT B complex is associated with the cell cycle. The hNat3 subunit of the hNatB complex has been found overexpressed in some forms of cancer.

Genetic determines activities of NAT that again regulate drug metabolism. Nearly 20% of Asians have an isozyme that results in slower N-acetylation of drugs, while 50% of Whites and African-Americans do.

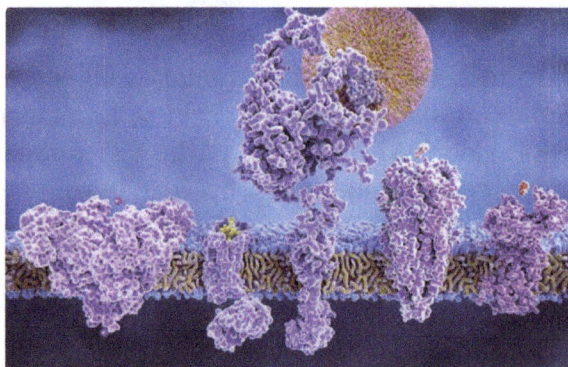

Different structures of membrane proteins: (left to right) Potassium channel, delta-opioid receptor, LDL receptor, acetylcholine receptor, histamine receptor, 3d rendering.

Lysine Acetylation and Deacetylation

The histone acetylation and deacetylation occurs on the lysine residues in the N-terminal tail as part of gene regulation. The mediating enzyme is often histone acetyltransferase (HAT) or histone deacetylase (HDAC). HATs and HDACs can modify the acetylation status of non-histone proteins as well.

Tubulin Acetylation

Tubulin acetylation and deacetylation has been studied in Chlamydomonas. A tubulin acetyltransferase located in the axoneme. It acetylates a specific lysine residue in the α-tubulin subunit in assembled microtubule.

Lipidation

Post-translational modification (PTM) of proteins includes the covalent addition of various lipids (e.g., fatty acids, isoprenoids, and cholesterol), which increases protein hydrophobicity to influence their localization and function. While this process is important to normal cell signaling, many proteins have gained attention as targets for lipid modifications due to misregulations observed in disease states. For example, Sonic hedgehog protein (Shh) is modified by cholesterylation before undergoing N-palmitoylation, which regulates signaling that is important for embryonic patterning and stem cell biology. The signaling protein Wnt can be modified by S-palmitoylation, which controls the transport of this ligand to its receptor during critical events in embryonic development, cell proliferation, and insulin sensitivity. The Ras superfamily of GTPases undergoes farnesylation, geranylgeranylation, and S-palmitoleoylation, which plays a central role in cancer development. Therapies targeting these modifications have the potential to impact a range of pathologies from cancer to viral infection. Therefore, developing specific tools to study them is essential. The most widely applicable are alkyne or azide-tagged lipid analogs that can be used in cell culture and in vitro studies as a chemical method for labeling and isolating lipidated proteins. Below we cover the basic biology of two of the most prominent lipidation modifications—acylation and prenylation.

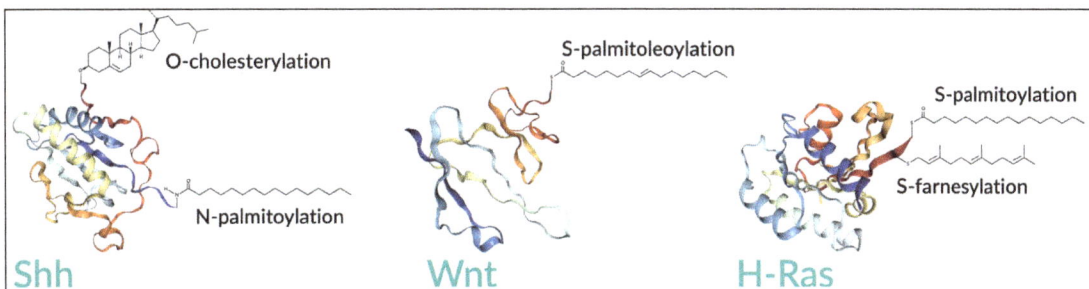

Protein Acylation

Protein acylation is catalyzed by acyltransferases. Palmitoylation, protein modification with palmitic acid, is the most common form of protein acylation, though other lipids, such as stearic acid or arachidonic acid, have been observed. Palmitoylation facilitates the association of proteins with cell membranes, mediates protein trafficking, and can regulate protein stability. S-Palmitoylation on cysteine residues is entirely reversible through the action of deacylases (depalmitoylases). The addition and removal of palmitate regulates protein distribution and function. In contrast to S-palmitoylation, N-palmitoylation is a stable modification of cysteine residues occurring at the protein's N-terminus.

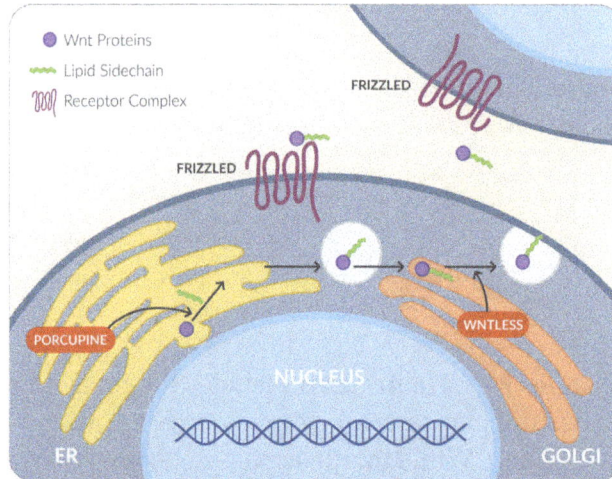

Wnt proteins are palmitoleoylated on a highly-conserved serine residue by porcupine (PORCN), a membrane-bound O-acyltransferase (MBOAT) that resides in the ER. Palmitoleoylation of Wnt proteins initiates their secretion and binding to the Frizzled receptor. In certain cancers with Wnt sensitivity, the inhibition of PORCN activity has become a therapeutic target for limiting Wnt secretion. The MBOAT, Hedgehog acyltransferase, N-palmitoylates Shh proteins prior to secretion from the endoplasmic reticulum, which is critical for their signaling range and efficacy.

Protein Prenylation

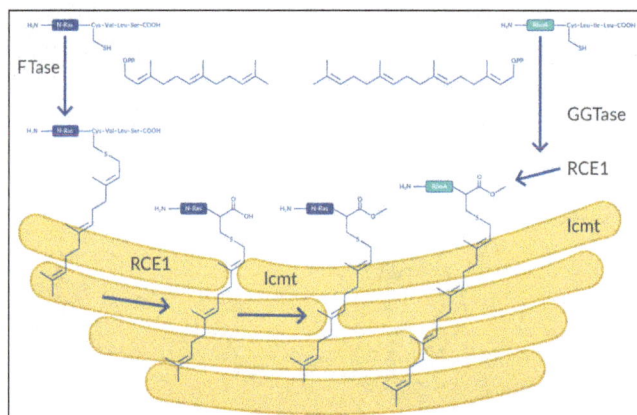

Prenylation is the enzymatic addition of either a farnesyl or a geranylgeranyl group to a C-terminal cysteine within the recognition sequence known as CaaX box. Proteins that undergo prenylation include those involved in cell cycle progression, oncogenesis, and parasitic infection. The process is catalyzed by three different enzymes: Farnesyltransferase (FTase), which adds a 15-carbon farnesyl group to proteins with the CaaX box, and two different geranylgeranyltransferases

(GGTase I and RabGGTase). GGTase adds 20-carbon geranylgeranyl groups to proteins with a CaaX sequence when X is leucine, and RabGGTase acts on Rab proteins, which do not have a CaaX box consensus sequence, with the assistance of a Rab escort protein. The isoprenoid addition drives proteins to associate with the endoplasmic reticular membrane where the converting enzyme RCE1 will cleave three terminal amino acids in order for isoprenylcysteine carboxyl methyltransferase (Icmt) to methylate the C-terminal prenylated cysteine. This sequence of events results in increased membrane affinity of the target proteins and is key for proper membrane association and protein-protein binding.

Proteolysis

Proteolysis is the hydrolysis of the peptide bonds that hold proteins together, resulting in the breakdown of proteins into their key components, peptides and amino acids. Proteolysis can occur as a method of regulation of cellular processes by reducing the concentration of a protein, transforming a protein into an active form, or by providing amino acids required to synthesize a different protein. Proteolysis is often performed by proteases, enzymes that catalyse the breakdown of proteins. It can also occur as a result of adverse cellular conditions such as extreme temperature, acidity, or salinity, which disrupts the molecules in the peptide bonds and results in the bonds breaking.

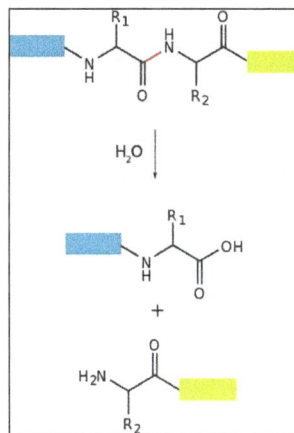

This figure depicts the breakdown of a peptide bond into the constituent amino acids through hydrolysis of the amide bond.

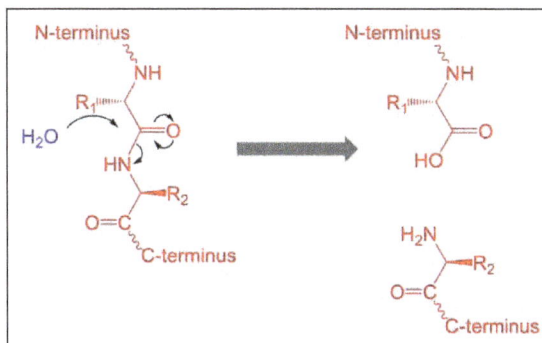

This figure depicts the hydrolysis of a peptide bond through the addition of water.

Proteolysis Structure

Proteolysis occurs when the peptide bonds holding a protein together are hydrolyzed. This often occurs through catalysis by proteases, enzymes that are involved in the breakdown of proteins. The enzymes interact with proteins with substrate specificity based on the conformation of the proteins and the amino acid residue to which they attach. Proteases function in one of two ways: They can break the amide bond between two amino acids at the amino or carboxy terminal (exopeptidases), or they can cleave the protein within the substrate (endopeptidases). Proteases can be categorized into six types, based on the way in which they hydrolyze proteins and the residue which is involved: Aspartate, cysteine, glutamate, metallo, serine, and threonine proteases. The different categories hydrolyze the amide bond either through an addition/elimination reaction that produces an intermediary, or by direct hydrolysis of the bond by a polarized water molecule. Proteases, and therefore proteolysis, can be inhibited by binding of inhibitors to the protease active sites, or through spatial confinement separating them from their substrates. This inhibition plays an important role in regulation of cellular processes.

Proteolysis Function

Proteolysis plays a number of important yet diverse roles in the body. It can control protein function either by producing or removing proteins. Proteolysis can regulate the concentration of a protein by removing any excess protein such that its function is diminished or impaired. Proteolysis is often responsible for complete inactivation of proteins, which can result in disruption of protein-protein interactions and signalling cascades including apoptosis. Alternatively, protcases can produce active proteins by conducting fine-scale modifications to a non-functioning or proto-protein, or by altering its physical state or location. Examples of this include regulation of blood clotting through the removal of clots by plasmin and fibrinolysis, and the activation of the protease trypsin by modification of the pre-protease zymogen.

Proteolysis also acts as cellular clean-up process by removing damaged or unnecessary proteins. Damaged proteins include those that have been fractured by another mechanism leaving only a protein fragment and misfolded proteins that are functionally impaired. The proteases will recycle the amino acids to make new or more appropriate proteins, or to relocate the proteins to a new region. Proteins that are intact and folded correctly will not be broken down, essentially acting as a cellular quality control function.

Proteolysis is a key component of nutrient digestion, breaking down any protein ingested so that the nutrients are available to be taken up by the organism. In this process the proteins are completely broken down into their amino acids. Proteolysis is also a component of food production, with proteases breaking down milk proteins in cheese production, and a low pH driving proteolysis of actin in dry sausage fermentation.

References

- Post-translational-modification, pierce-protein-methods, protein-biology-resource, life-science: thermofisher.com, Retrieved 15 February, 2019

- Ubiquitination, life-sciences: news-medical.net, Retrieved 17 June, 2019

- Phosphorylation, pierce-protein-methods, protein-biology, life-science: thermofisher.com, Retrieved 27 January, 2019

- Methods-detecting-protein-phosphorylation: rndsystems.com, Retrieved 22 May, 2019

- Protein-glycosylation, pierce-protein-methods, protein-biology, life-science: thermofisher.com, Retrieved 1 March, 2019

- S-nitrosylation: creative-proteomics.com, Retrieved 19 July, 2019

- Protein-methylation: biosyn.com, Retrieved 9 January, 2019

- Acetylation-of-Proteins: news-medical.net, Retrieved 27 April, 2019

- Post-translational-modification, protein-lipidation: caymanchem.com, Retrieved 7 August, 2019

- Proteolysis: biologydictionary.net, Retrieved 28 May, 2019

Chapter 5

Applications of Proteomics

Proteomics involves the large-scale experimental analysis of proteins and proteomes. It has helped in the identification of a large number of proteins. Proteomics is applied in a variety of fields such as biotechnology, environment, medicine, food industry etc. The diverse applications of proteomics in these different areas has been thoroughly discussed in this chapter.

Proteomics in Biotechnology

Proteomics is a nascent technology with much to develop and much to offer. Both the biotechnology and pharmaceutical arenas will be highly impacted by the new window proteomics is opening. This area of studies will permit us to identify, characterize, and quantify proteins on a massive scale, resulting in a fundamental change in biological and chemical approaches to understanding proteins and even the ability to address new important questions. Advancing the tools of proteomic technology further is of major interest to biotech and pharma communities. Proteomics does require substantial resources and infrastructure investment; therefore, the effort will be greatly enhanced by the coordinated participation of all involved, in both academia and industry. Such resources and coordination are only slowly becoming available on the scale necessary to make a significant impact and the Human Proteome Organization (HUPO) is at the vanguard of these efforts.

Now that the human genome has been sequenced, the study of the proteins that carry out the gene's instruction is likely to be the next frontier in biomedical research. Surprisingly, the human genome appears to contain perhaps fewer than 30 000 genes, which is far fewer than the original estimates. Enter the world of proteomics. Proteomics includes not only the identification and location of proteins, but also the determination of networks, interactions, activities, and functions. A single gene can encode multiple proteins through a variety of translational and post-translational mechanisms. In addition, the same protein can assume multiple forms. Finally, the same protein may have multiple functions, depending on its cellular environment and associated networks. All of these possibilities result in a proteome (the ensemble of proteins related to a genome) estimated to be at least an order of magnitude more complex than the genome itself.

Experimental studies related to protein fractionation and separation, qualitative and quantitative protein analyses, identification/characterization of the composition of protein complexes and organelles seem likely to enhance our understanding of dynamic cellular processes, thus, making the proteome a less formidable challenge. Looking ahead we expect protein arrays to play a powerful important role and that novel protein therapeutics will emerge.

Protein Production

The challenge of studying proteins in a global way is driving development of new technologies for systematic and comprehensive analysis of protein structure and function. Experimental uses of proteins for structural and functional studies typically require milligram amounts in purified form. Protein expression, production, and purification are fundamental processes in these studies; however, they have typically only been applied on a case-by-case basis to proteins of interest. Strategies are emerging to industrialize these processes and overcome the limitations imposed by conventional methods. Parallel processing, miniaturization approaches, array spotting techniques all will significantly contribute to making proteomics more main stream.

Antibodies

To define the proteome of an organism, there is a need for robust reproducible methods for the quantitative detection of all the polypeptides in a cell. Because of their high-affinity, specificity, and their ability to bind virtually to any protein, antibodies appear particularly promising as the receptor element in protein-detection arrays. For proteomic-scale analyses and fabrication of high-density arrays, the ability to produce and isolate antibodies en masse to a large number of target molecules is critical. The high-throughput issues of recombinant proteins apply equally well to antibody molecules to generate proteins for microarrays and antibodies for immunolocalization and/or functional analysis utilizing highly sensitive detection protocols.

Environmental Applications of Proteomics

The application of proteomics in studies of microbial physiology, metabolism, and ecology in the context of natural and engineered soil and water environments are described. These habitats often contain a very diverse population (e.g., ca. 104 prokaryotic species in 30 cm^3 of forest soil) with total population sizes that vary over many orders of magnitude. Despite a growing knowledge of the range of microbial diversity, most of the microorganisms seen in natural environments are uncultivated, and their functional roles and interactions are unknown. The metabolic capabilities of microorganisms, including dehalogenation, methanogenesis, denitrification and sulfate reduction, are studied for their applications in environmental biotechnology. In addition, the abilities of some microorganisms to tolerate radiation and toxic chemicals, to use many different electron donors and acceptors, or to survive at extremes of environmental conditions are all of interest. Furthermore, microorganisms in both natural and engineered environments generally function in communities, allowing them to benefit from syntrophism, exchange of genes and cell–cell communication, among other phenomena. However, few details are known about these interactions. Similarly, little is known about how naturally occurring microbial communities respond to perturbations such as starvation, desiccation, or freeze-thaw cycles.

Using proteomics, one can determine protein expression profiles related to these research questions for both microbial isolates and communities. Proteomics provides a global view of the protein complement of biological systems and, in combination with other omics technologies, has an important role in helping uncover the mechanisms of these cellular processes and thereby advance the development of environmental biotechnologies.

As a field, environmental proteomics is much less developed than other proteomics applications areas. Most published proteomics studies focus on one organism or cell type, and the effects of the growth environment are investigated by comparing different controlled conditions. One challenge of environmental proteomics is that the environment of interest is not controlled, and is difficult to emulate in the laboratory. Furthermore, issues related to uncultured and/or unsequenced organisms and protein extraction from native samples are key to the success of environmental proteomics studies. An important recent advance in environmental proteomics is the ability to identify proteins from unsequenced organisms with the use of modern bioinformatics techniques. Cross-species protein identification and protein sequence similarity searches are the most common strategies used to identify proteins when the genomic sequence is not available. However, caution must be used since these approaches can have low rates of success and require careful statistical analysis in order to avoid false positive identifications.

An important recent development in environmental proteomics that introduces new promises and challenges is the analysis of the collective proteome of microbial communities, known as metaproteomics. Here, the community is viewed as a 'metaorganism', in which population and meta-proteome shifts are forms of functional responses. This approach has been used by a few research groups and has shown great potential in the evaluation of biological processes in a community without isolating organisms. It also allows for a view of organism interactions, which are impossible to determine using pure cultures. Metaproteomic samples are biologically highly complex, which makes these studies especially challenging. If one considers that a typical bacterium contains approximately 3000 genes then a metaorganism constituted by 100 species would have about 3×10^5 genes and a proteome of corresponding complexity. Some of the main challenges in metaproteomics are the difficulties related to evaluating such a large number of gene products as well as the lack of genome sequences for the large majority of environmental bacteria. Nonetheless, important progress has been made with fascinating results, including advances that allowed for the extraction and identification of proteins directly from soil or seawater.

Environmental proteomics, including metaproteomics, yields better results in combination with other omics approaches such as metabolomics and transcriptomics. In addition, proteomics allows one to confirm the existence of gene products predicted from a DNA sequence, providing a major contribution to genomic science and an effective complement to nucleic-acid-based methods as a problem-solving tool in molecular biology. In addition, proteomics can be used for phylogenetic classification of bacterial species, either by using 2D maps or peptide sequences obtained from mass spectrometry. Proteomics has the advantage of not being limited to organisms for which the genomic sequence is available. In addition, the proteome represents the actual enzyme content in a system, going beyond potential gene expression as determined by microarrays, and can provide information about post-translational modifications. Another technique of great potential, especially when combined with metaproteomics, is the recently developed pyrosequencing approach, which has already been applied in some metagenomics projects. Current proteomics methods have limitations; for example, it is not yet possible to acquire data on all the proteins present in a sample, due mainly to their large concentration range and the lack of a method for amplifying low-abundance proteins. Thus, it is often advantageous to complement proteomics with other omics tools.

Wastewater Treatment

Protein profiling related to wastewater treatment has primarily used SDS–PAGE (1D electrophoresis, 1DE) to characterize the organisms involved in this process and their ecology. Such studies have focused on the diversity of organisms in a treatment system and the influence of environment on protein profiles rather than the identification of interactions or specific metabolic pathways. For example, MacRae and Smit described 33 different strains of *Caulobacter* present in wastewater. The strains were distinguished based on colony characteristics, DNA, and protein profiles using 1DE. They also point out the increasing antibiotic resistance of these strains, indicating environmental adaptation. Jacob *et al.* used 1DE to characterize 24 different strains of *Campylobacter* present in a wastewater treatment plant. Their work detected the presence of different strains, based on evidence of different protein band patterns. A similar study was conducted by Niemi *et al.* and involved 371 environmental isolates of fecal streptococci samples. Samples were collected from domestic and industrial wastewater, and were characterized and clustered into seven groups according to their 1DE protein profiles. Samples from each environment had typical species compositions, and their protein profiles varied according to their environments. Maszenan *et al.* performed a similar analysis for strains of the species *Acinetobacter*, together with a range of isolates from a biological nutrient-removal activated sludge plant. These studies show that the idea of studying the proteome of a community of organisms, i.e. metaproteomics, has been in development by environmental researchers for at least two decades, even though the laboratory and bioinformatic methodologies available were limiting.

Wagner-Dobler *et al.* pursued the goal of better understanding bacterial communities capable of degrading biphenyl for future bioremediation applications. Different species were identified using 16S rDNA methods, and 1DE of whole-cell proteins was used to provide information on the similarity to strains of the same species. Comparison of normalized protein patterns revealed that all of the representative isolates were very similar to each other, thus likely coming from the same species. More recently, Francisco *et al.* studied the proteome of a microbial community under chronic chromate stress in an effort to better understand and improve microbial metal remediation of a chromium-contaminated activated sludge. Using numerical analysis of protein patterns and correlating these with lipid profiles, they were able to cluster the organisms in the community into subgroups that shared similar metabolic abilities. The main findings here were that the protein and lipid clusters were in good agreement and that, within the same protein and lipid cluster, there were functional differences in the chromium resistance and reducing abilities of the strains in the community.

Metabolic Engineering

Although not native to soil and water environments, *E. coli* has been studied in the environmental context because of its role as a platform for metabolic engineering. Pferdeort *et al.* investigated the proteome of *E. coli* metabolically engineered for trichloroethene biodegradation by the introduction of six genes of an evolved toluene ortho-monooxygenase from *Burkholderia cepacia G4*. The cellular physiology of the engineered strain was significantly altered due to the insertion of the toluene ortho-monooxygenase genes, with differential regulation of 45 proteins. Another study by Lee *et al.* analyzed strains from the next stage of the metabolic engineered strategy in which protective enzymes (glutathione S-transferase or epoxide hydrolase) were inserted. Using a quantitative proteomics approach, they found that some of the induced proteins were involved in the oxidative defense mechanism, pyruvate metabolism and glutathione synthesis. Proteins involved in indole synthesis, fatty acid synthesis, gluconeogenesis and the tricarboxylic acid cycle were repressed. Proteomic studies of the effects of metabolic engineering are essential for the identification and quantification of the changes in host cell physiology reflected by protein production or other cellular processes. Since most bacterial cellular processes are either regulated or directly carried out by proteins or protein complexes, physiological responses to new genes can be expected to result in altered production of various host cell proteins other than those introduced in the genetic manipulation.

Microbial Ecology

Ecological studies focus on naturally occurring bacterial adaptation to their environments. Proteomics has been used in several studies to provide insights into the mechanisms of adaptation, especially to extremes of temperature. Proteins of hyperthermophilic organisms are of particular importance since they have an enhanced conformational stability, allowing them to be active at high temperatures. This property can be used to investigate the molecular basis of protein folding and conformational stability. Prosinecki *et al.* studied hyperstable proteins from *Sulfurispharea* sp., a hyperthermophilic archaeon that is able to grow between 70 °C and 97 °C. They dynamically perturbed the proteome and identified proteins with enhanced stabilities, involved in key cellular processes such as detoxification, nucleic-acid processing and energy metabolism. These proteins were still biologically active after extensive thermal treatment of the proteome.

Other ecological studies have focused on cold adaptation of bacteria. Proteomic analysis was used by Qiu *et al.* to investigate the cold adaptation of *Exiguobacterium sibiricum* 255-15, a strain isolated from Siberian permafrost sediment. They used an alternative approach involving chromatofocusing coupled to mass spectrometry to identify 256 proteins preferentially or uniquely expressed at 4 °C. Among these were 39 cold acclimation proteins, including chaperones, and three cold shock proteins. These results indicated that the adaptive nature of *E. sibiricum* 255-15 at near-freezing temperatures could be regulated by cellular physiological processes through the regulation of specific cellular proteins. The researchers concluded that the proteins that were up-regulated at the lower temperatures may enable the cells to adapt to near or below-freezing temperatures. Here it was shown that in order to understand the biological context of bacterial cold adaptation, large-scale proteomic studies are necessary to uncover all cellular processes and not only small sets of proteins isolated in specific functional contexts. Methé *et al.* in a similar study, used *Colwellia psychrerythraea* 34H and found changes to the cell membrane fluidity, uptake and synthesis of cryotolerance compounds, and strategies to overcome temperature-dependent barriers to carbon uptake. The salt and cold adaptation of *Psychrobacter* 273-4 was evaluated by Zheng *et al.* Different proteins were identified in cold adaptation in the presence of salt, showing a combination effect of salt and cold on protein expression.

Environmental Stress Responses

Proteomics approaches have often been used to gain insights into the physiological responses of microorganisms to temperature, chemical and other stresses. The choice of proteomics as the primary experimental tool is a reflection of the ability to obtain system-wide information for non-model organisms (e.g. given the cost of procuring DNA microarrays for these species) and to obtain protein identifications without a genomic sequence. Thus, both 2DE- and chromatography-based proteomics methods have been used to investigate the mechanisms of tolerance to such stresses as low temperatures, high temperatures, acidic conditions, organic solvents, heavy metals and oxidizing chemicals. While the up-regulation of known stress-response proteins was frequently observed in these studies, there were also discoveries of proteins involved in other detoxification or adaptation strategies, including novel transporter proteins, lipid biosynthesis pathways and osmoprotectants. Moreover, the regulation of stress responses could be discerned, particularly when the same species was exposed to different stresses (e.g. nitrate, salinity and high temperature for *D. vulgaris*). In the response of *D. vulgaris* Hildenborough to growth inhibitory levels of nitrate stress, it was found that proteins involved in central metabolism and sulfate reduction were unaffected. However, up-regulation was observed in nitrate reduction systems, transport systems for proline, glycine-betaine and glutamate, oxidative stress proteins, ABC transport systems as well as in iron-sulphur-cluster-containing proteins. In the case of increased salinity, *D. vulgaris* responded with up-regulated efflux systems, ATPases, RNA and DNA helicases, and chemotaxis genes. Down-regulated systems included flagellar biosynthesis, lactate uptake permeases and ABC transport systems. These results demonstrated that *D. vulgaris* responded similarly to NaCl and KCl stresses. In the case of the response of *D. vulgaris* to heat shock, proteomic analysis revealed the up-regulation of heat shock proteins, protein turnover and chaperones, and down-regulation of energy production and conversion, nucleotide transport, metabolism, translation and ribosomal structure. The proteomics study also suggested the possibility of posttranslational modifications in the chaperones and in several periplasmic ABC transporters. It is clear from this set of studies that proteomic analysis not only reveals system-wide stress responses but also has the ability to identify specific mechanisms of defense that characterize each stress condition.

Application of Proteomics in Food Industry

Food and human nutrition make an important bio-mixture; therefore their quality control and safety are very essential. Since proteins are the main constituents of foods, proteomic technology can monitor and characterize the protein content of foods and their changes during production using two-dimensional polyacrylamide gel electrophoresis (2D-PAGE) and chromatography techniques in combination with mass-spectrometry. Proteomics technology can be used for detection, validation, optimization, and also quality control of food industry. Concerning food science, food-proteomics methods can help to identify quality biomarkers to design better and safer foods. Therefore, proteomics may help food-producers to provide foods which give more guarantees of human health and safety. Thus, modern nutritional research focuses on health promotion, performance improvement, disease prevention, protection against toxicity, and stress. Some of the side effects of foods are due to contamination with microorganisms or their corresponding endo/exotoxins. Several numbers of human hospitalizations and even deaths happen because of microbial contaminations each year. Proteomics is a useful tool for identification of microbial contaminations and their toxins. In some cases, food products with animal origin such as seafood and milk create allergic reactions; in this regard, proteomics can be a very important tool for detection of allergens origin. Proteomic techniques are increasingly used for quality control and safety of raw materials in food industry as well. Most of the proteomic researches in the food technology are performed by the use of comparative 2D-PAGE and quantitative proteomics. Figure shows proteomics tools in the processing of food production and quality controls. Proteomics applications can be divided into several groups in food investigation including cereal grains, fruits, eggs, meat products, and seafood.

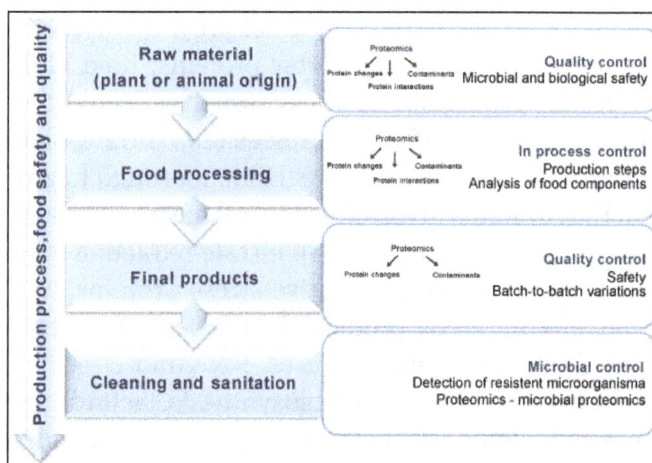

Use of proteomics in the development pathway for food production,
and assessing food safety, originality and quality.

With rapidly growing global population, climate changes and increasing need for natural resources, food production with efficient crop yield is critical. To achieve these aims, novel methodology for vitality of seeds and protecting crops against stress are required. "Omics" technologies are promising tools for such objectives. A specific advantage of proteomics over other "Omics" techniques is the capacity to determine the functional impact of protein modifications on crop plant productivity. The application of proteomics for analyzing crop plants increased during the last decade. Proteomics studies have identified numerous proteins that play crucial roles in plant growth and development. Seeds are one of the most important factors in crop production, as crop yield is related to seed vitality. One of the traditional breeding methods is transgenic technique to obtain crops with desired qualities. Evaluation of these genetically modified crops with proteomic methods is essential. Another aspect of proteomic studies in agriculture is related to interaction between crops and other organisms that influence the growth yield of crops. Proteomic analysis can complement molecular genetics approaches for studying the mechanisms by which pathogens attack cereal crops. Also proteomics approach is an efficient tool to analyze agriculture crop biomass. Systems biology analysis will also help the breeding of robust crop plants that are tolerant to environmental stresses and have high nutritional value.

Medical Application of Proteomics

Proteomics in Medicine

Proteomics is vital for decrypting how proteins interact as a system and for comprehending the functions of cellular systems in human disease. Nevertheless, due the fact that the proteome is several orders of magnitude more complex than the genome and highly fluid in nature, large-scale proteomic analysis remains challenging.

Cancer biologists have made the first attempts to utilize proteomics for diagnostic and prognostic purposes. A serum-based proteomic pattern diagnostics has soon been developed, which represents a new method of diagnosis and disease identification for ovarian cancer detection.

The concept behind this is that the diagnostic endpoint for ovarian cancer detection is not a single analyte, but a proteomic pattern composed of many individual proteins. Furthermore, defining signaling pathways in ovarian cancer cells through proteomic analysis gives us the opportunity to optimize the use of molecularly targeted agents against central and biologically active pathways.

Protein-sequence data are now available for many microorganisms, providing us with tools for understanding their resistance to antimicrobial drugs and for identifying novel agents for treating drug-resistant disease. Surface-enhanced laser desorption/ionization-time of flight (SELDI-TOF) is now employed to rapidly diagnose invasive aspergillosis, tuberculosis, sleeping sickness and Chagas' disease.

Advances in mass spectrometry, coupled with better isolation and enrichment techniques which allows the separation of organelles and membrane proteins, made the in-depth analysis of cardiac proteome a reality. Evolution of proteomic techniques has allowed more detailed investigation of molecular mechanisms underlying cardiovascular disease, facilitating the identification of not only modified proteins, but also the nature of their modification.

Proteomics is also becoming a part of the quality-control process in transfusion medicine with an aim to verify the identity, safety, potency and purity of various blood products. The proteomic approach is a valuable way to implement a global screening of storage-related lesions of red blood cells and to study the mechanisms of possible biological consequences on the transfusion recipient.

Proteomics in Drug Development

Since disease processes and treatments are often manifesting at the protein level, proteomics has received much attention as a drug development platform. The pattern of protein changes after drug application gives important information about the mechanism of action, either for therapeutic or toxicological effects.

A majority of large pharmaceutical companies now have a proteomics-oriented biotechnological (or academic) partner or have started their own proteomics division. Usual applications of this field in the drug industry include target identification and validation, identification of biomarker efficacy and toxicity from easily attainable biological fluids, as well as investigations into mechanisms of drug action or toxicity.

Proteome mining is today used to discover new antimalarial drugs that target purine binding proteins in the blood stage of infection. It represents a functional proteomics approach used to assess protein information from the analysis of specific subproteomes. This approach exploits the serendipitous nature of drug discovery, simply because it expands the hit rate over a conventional screen by a factorial of the proteome that is bound.

Many of the top-selling drugs today either act by targeting proteins or are proteins themselves. Advances in proteomics may help researchers to eventually create medications that are "personalized" for different individuals in order to achieve better effectiveness and fewer side effects.

Proteomics Application in Biomarker discovery

A biomarker usually refers to disease- related proteins or a biochemical indicator that can be used in the clinic to diagnose or monitor the activity of disease, prognosis, and development of the disease, and also to guide the molecular target treatment or evaluation of the therapeutic response. In medicine, a biomarker can be a traceable substance that is introduced into an organism as a means to examine organ function or other aspects of health. One example of a commonly used biomarker in medicine is PSA (prostate specific antigen). Nevertheless, malignancies are usually detected at severe stages when patients have very poor prognosis and few treatment options are present which are mostly due to a high cost and time -consuming process in biomarker tracing. Furthermore, development of a better throughput analyzing method is a critical requirement for early detection; and combination of various platforms of onco-proteome data is required. As protein expression alters during disease condition in biological pathways, monitoring of these altered proteins in tissue, blood, urine, or other biological samples can provide indicators for the disease. Proteomics technology has been extensively used in the molecular medicine especially for biomarker discovery. By analyzing of a global protein profiling in the body fluids, proteomics can identify invaluable disease-specific biomarkers. Expression of proteomics provides biomarker detection through comparison of protein expression profile between normal samples vs. disease affected ones. The simplest approach used in biomarker discovery is 2D-PAGE in which protein profiles are compared between normal and disease samples such as tumor tissues and body fluids. Disease-specific biomarkers can be divided into diagnostic, prognostic, and treatment predictive biomarkers according to information which they provide. A diagnostic biomarker is used for early detection or presence of the disease. A prognostic biomarker usually is used to predict the recurrence and aggression of disease and a patient response to treatment by a given drug. Predictive biomarkers are useful tools to classify the patients into responder and non-responder groups. This classification also is important in drug design applications. According to estimates, only 2% of human diseases appear to be due to a single gene damage, and epigenetic and environmental factors involved in the development and outcome of the diseases account for remaining 98%. In this regard, proteomics can be helpful to identification of proteins that can potentially serve as disease-associated biomarkers which involved in disease progression. After identifying biomarkers by mass spectrometry-based approach, biomarkers need to process using bioinformatics analyses and also need to be reproduced in different populations. Despite amplifying the interest, high investigation burdens, and high level of publications, unfortunately, a few of identified biomarkers using proteomics technology have been validated and approved by FDA for clinical usage yet.

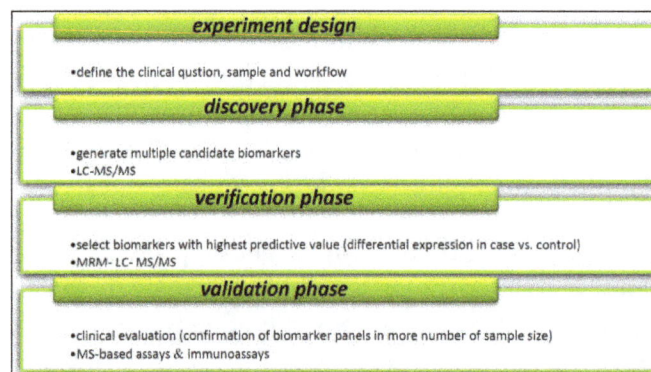

experiment design
• define the clinical qustion, sample and workflow

discovery phase
• generate multiple candidate biomarkers
• LC-MS/MS

verification phase
• select biomarkers with highest predictive value (differential expression in case vs. control)
• MRM- LC- MS/MS

validation phase
• clinical evaluation (confirmation of biomarker panels in more number of sample size)
• MS-based assays & immunoassays

Different steps of new biomarker development. Majority of methods for analysis of disease-specific biomarkers are based on mass spectrometry (MS). Variety of separation methods including liquid chromatography (LC), electrophoresis (E), two-dimensional electrophoresis-MS

(2DE-MS), 2D-polyacrilamid gel electrophoresis-MS (2D-PAGE-MS), matrix- assisted laser desorption/ionization- time of flight-MS (MALDI-TOF-MS), surface- enhanced laser desorption/ ionization-TOF-MS (SELDI-TOF-MS), LC-MS/MS, Fourier transform ion cyclotron resonance-MS (FTICR-MS), multiple reaction monitoring/ selected reaction monitoring (MRM/SRM) in combination with MS use in discovery step of biomarker identification process. In validation step, several techniques such as enzyme- linked immunosorbent assay (ELISA), arrays, MRM/SRM, western blot (WB) and immune histochemistry (IHC) can be used.

Concerning biomarker discovery, proteomics technology is a promising tool for disease-associated biomarker detection in the biological fluids including urine, plasma, serum, etc. It is very important that body fluid samplings for proteomics research are less invasive and have low-cost advantages. The proteomics biomarker discovery is advanced in a variety of diseases such as cancer, cardiovascular diseases, acquired immune deficiency syndrome (AIDS), renal diseases, diabetes, etc. However, despite many technological developments, some challenges remain to be overcome as body fluids are highly complex mixtures of proteins and contain high dynamic range of proteins. Each of these sample types can be used in different disease conditions, for example in kidney disease; urine is a valuable sample because of urine proteins directly reflect changes in the kidney function. Blood is a logical fluid to be used for biomarker discovery in human disease, since it has several merits over other samples presented in Figure. On the other hand, biomarker identification using plasma proteomics has challenges including (i) high dynamic range of plasma proteins, (ii) low abundance of invaluable biomarkers in plasma, and (iii) patients' variation. Despite the advancement of proteomic methods, the current single method cannot simultaneously conquer the above challenges in biomarker discovery process in plasma proteome yet. 2D-PAGE/MALDI-TOF and surface-enhanced laser desorption/ionization (SELDI)/Protein Chip techniques are the most approaches used to identify biomarkers for different disease. SELDI-MS is used for biomarker detection in a variety of diseases due to its ease of operation and high-throughput application.

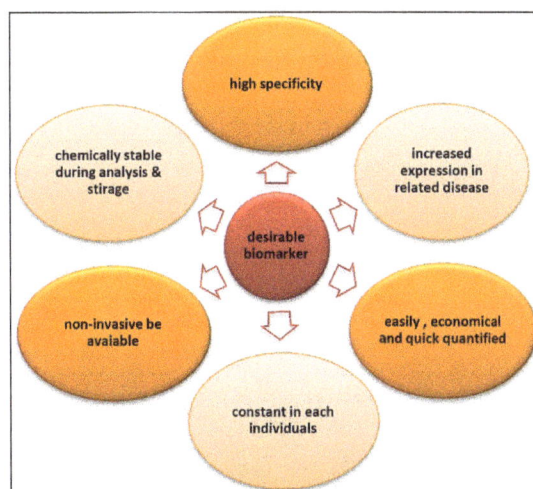

Ideal biomarkers features. The ideal biomarker should have high specificity for a certain disease condition. Proteomics technology is powerful tool for biomarker discovery through characterization and evaluation of global profiling of proteins under given state.

Due to cancer importance, the most proteomics studies in the field of biomarker discovery concerned cancer diseases. Proteomics as one of the modern areas of biochemistry holds great promise in the cancer study. Since proteome represents actual state of the cell, tissue, or organism, there are suitable biomarkers related to the tumors which can be used for diagnostic proposes or follow up of patients. It should be noted that one of the major challenges facing the biomarker discovery especially in cancer is heterogeneity between patients. Therefore, personal medicine has become a popular trend for cancers. The combined use of biomarkers as biomarker panels has also been shown to give a better prognosis and increase sensitivity and specificity to predict the response of patients to therapy. Since advantages of biomarker panels (or biomarker patterns) have been confirmed in several publications recently, in this context proteomics can be useful tool for identification and verification of such biomarkers and panels.

Permissions

Index

A

Affinity Chromatography, 69, 71-72, 83, 95, 120-121, 123-129, 161, 174, 200

Arginine, 5-6, 8-9, 14-16, 21, 23, 30, 37, 55, 135, 218

Asparagine, 4-5, 14, 16, 55, 208, 210-212

Aspartic Acid, 4-5, 15-16, 30, 36, 55, 61, 89

Avidin, 29, 200

B

Bimolecular Fluorescence Complementation, 178

C

Ceruloplasmin, 28, 32

Co-immunoprecipitation, 157, 160-162, 165-166, 174, 183, 189, 191-192

Collagen, 5, 7, 11, 22-25, 31, 215

Conjugated Proteins, 23, 28, 30-31

Contractile Protein, 13, 23, 26

Cysteine, 4, 17, 20, 32, 36, 48, 50, 55, 60-61, 71, 77, 89, 105, 172, 195, 217, 222-224

D

Dna-binding Proteins, 56, 127, 218

E

Environmental Proteomics, 227-228

Enzyme-linked Immunosorbent Assay, 69, 72, 203

F

Fibrinogen, 13, 27, 47

Fibrous Proteins, 9, 13-14, 17, 23

Fusion Tag, 175

G

Gel Electrophoresis, 16, 62-63, 66, 69, 74-76, 93, 98-100, 102, 104-105, 177, 189, 201, 231, 235

Globulins, 11, 13-14, 22-23, 27-28, 30, 38

Glutamine, 5, 14, 16, 30, 55

Glycine, 3-5, 7-9, 15-16, 19, 24-25, 55, 100-101, 151, 162, 164, 172, 189, 195, 231

Glycoprotein, 28-29, 31, 77, 207-210, 213-214

H

Heme Protein, 35

Histidine, 5-6, 8, 15-16, 21, 30, 32-34, 52, 72, 88, 128, 130-131, 175, 218

Histones, 23, 30, 33, 127, 218-220

Hydrochloric Acid, 8, 15, 38

Hydrophobic Interaction Chromatography, 94, 113-114, 117

I

Immunoprecipitation, 83, 96, 106, 157, 160-162, 165-166, 174, 183, 189, 191-192, 199

in Vitro Crosslinking, 186-188

Isoleucine, 3-5, 10, 16, 20, 36-37, 55, 113, 139

K

Keratin, 5, 23, 25

L

Lectins, 72, 95, 126, 209, 213

Leucine, 3-5, 10, 16, 20, 37, 39, 113, 158-159, 172, 223

Lipoproteins, 23, 28-29, 31, 127

Lysine, 5-9, 14-16, 25, 30, 48, 55, 60, 135, 195, 215, 218-219, 221

M

Mass Spectrometry, 62-63, 66, 69, 77-80, 82-84, 105-107, 120, 160-162, 189-190, 199, 201, 206, 210-211, 228, 230, 233-234

Metalloproteins, 31-32, 99

Metaproteomics, 228-229

Methionine, 4-5, 16, 55, 219

Mucoprotein, 28, 31, 37

Myosin, 5, 13, 23, 26-27

N

Nuclear Magnetic Resonance, 69, 135

O

Ovalbumin, 13, 29

P

Peptide, 1, 3, 5, 8-10, 12-15, 17-21, , 31, 34, 37, 39, 53, 56-58, 61, 63, 67, 72, 80, 93, 96, 105, 107, 118, 128, 132, 134, 138, 143, 148, 162, 170, 190, 194, 205, 210, 218, 228

Phage Display, 72, 169-173, 192

Phenylalanine, 5-8, 10, 16, 19-20, 37, 42, 55, 113

Phosphoprotein, 29, 199-200

Polypeptide Chains, 2, 220

Post-translational Modification, 80, 83, 194, 196, 206-207, 211-212, 215-216, 220-221

Prosthetic Group, 23, 30-32, 35, 44

Protamines, 6, 30, 33

Protease Inhibitors, 88-91

Proteases, 54, 58, 61, 71, 76, 82, 88-89, 91, 126-127, 194, 223-224

Protein Complex, 32, 160-162, 164-165, 176-177

Protein Denaturation, 20, 50, 58, 116, 119, 125

Protein Expression, 67, 69, 74, 82, 102, 157, 167, 175, 189, 204, 227, 230, 234

Protein Fractionation, 133, 226

Protein Kinases, 74, 197-198

Protein Microarrays, 69, 73-74, 166-168

Protein Phosphorylation, 80, 84, 195-201, 203, 206

Protein Profiling, 74-76, 79, 81, 206, 228, 234

Protein Prote in Interactions, 157

Protein Purification, 71, 86, 88-89, 91-95, 97, 99, 108, 111-112, 125, 128, 133, 174

Protein Sequence, 89, 157-158, 228

Protein-dna Interactions, 170, 182

Proteolysis, 58, 61, 89, 91, 126, 165, 193, 199, 223-225

Proteome Analysis, 69-70, 74, 76-78, 99

Pull-down Assay, 174-178

Q

Quaternary Structure, 9-10

S

Scalar Coupling, 141, 146, 153

Serum Albumin, 13, 28-29

Silver Staining, 101

Size Exclusion Chromatography, 69, 71, 93-94, 98, 109, 131

T

Threonine, 4-5, 14, 16, 55, 195-197, 199-200, 208, 215-216, 224

Tripeptide, 3, 73

Trypsin, 8, 21, 24, 26, 42, 44, 58, 61, 66, 127, 135, 224

U

Ubiquitination, 157, 193-195, 224

V

Valine, 3-5, 10, 12, 16, 20, 37, 39, 55, 113, 143-144, 151

W

Western Blot, 73, 102-103, 157, 160-162, 164-165, 175-176, 189, 200, 202-203, 235

X

X-ray Diffraction, 12-13, 17, 19

Y

Yeast Two-hybrid Assay, 63

www.ingramcontent.com/pod-product-compliance
Lightning Source LLC
Chambersburg PA
CBHW061257190326
41458CB00011B/3693